Lecture Notes in Economics and Mathematical Systems

669

Founding Editors:

M. Beckmann
H.P. Künzi

Managing Editors:

Prof. Dr. G. Fandel
Fachbereich Wirtschaftswissenschaften
Fernuniversität Hagen
Hagen, Germany

Prof. Dr. W. Trockel
Murat Sertel Institute for Advanced Economic Research
Istanbul Bilgi University
Istanbul, Turkey

and

Institut für Mathematische Wirtschaftsforschung (IMW)
Universität Bielefeld
Bielefeld, Germany

Editorial Board:

H. Dawid, D. Dimitrov, A. Gerber, C-J. Haake, C. Hofmann, T. Pfeiffer,
R. Slowiński, W.H.M. Zijm

For further volumes:
http://www.springer.com/series/300

Stephan Leitner • Friederike Wall
Editors

Artificial Economics and Self Organization

Agent-Based Approaches to Economics and Social Systems

 Springer

Editors
Stephan Leitner
Friederike Wall
Department for Controlling and Strategic
  Management
Alpen-Adria-Universität Klagenfurt
Klagenfurt
Austria

ISSN 0075-8442
ISBN 978-3-319-00911-7   ISBN 978-3-319-00912-4 (eBook)
DOI 10.1007/978-3-319-00912-4
Springer Cham Heidelberg New York Dordrecht London

Library of Congress Control Number: 2013946880

© Springer International Publishing Switzerland 2014
This work is subject to copyright. All rights are reserved by the Publisher, whether the whole or part of the material is concerned, specifically the rights of translation, reprinting, reuse of illustrations, recitation, broadcasting, reproduction on microfilms or in any other physical way, and transmission or information storage and retrieval, electronic adaptation, computer software, or by similar or dissimilar methodology now known or hereafter developed. Exempted from this legal reservation are brief excerpts in connection with reviews or scholarly analysis or material supplied specifically for the purpose of being entered and executed on a computer system, for exclusive use by the purchaser of the work. Duplication of this publication or parts thereof is permitted only under the provisions of the Copyright Law of the Publisher's location, in its current version, and permission for use must always be obtained from Springer. Permissions for use may be obtained through RightsLink at the Copyright Clearance Center. Violations are liable to prosecution under the respective Copyright Law.
The use of general descriptive names, registered names, trademarks, service marks, etc. in this publication does not imply, even in the absence of a specific statement, that such names are exempt from the relevant protective laws and regulations and therefore free for general use.
While the advice and information in this book are believed to be true and accurate at the date of publication, neither the authors nor the editors nor the publisher can accept any legal responsibility for any errors or omissions that may be made. The publisher makes no warranty, express or implied, with respect to the material contained herein.

Printed on acid-free paper

Springer is part of Springer Science+Business Media (www.springer.com)

Preface

In this year, the Artificial Economics conference reaches its 9th anniversary, and the aims and topics still attract researchers to contribute to this symposium. Since 2005, the Artificial Economics conferences bring together researchers from computer science and economics and encourages multi-disciplinary research in economics. Moreover, it appears that this conference series enfolds the potential of developing a growing community in this exiting field of research.

Two features might be regarded as the main building blocks of the Artificial Economics: agent-based models and the use of computational techniques to "solve" them. In particular, artificial markets or social systems, artificial networks or artificial organizations consisting of interacting, heterogeneous agents are modeled, computationally represented and simulated. The behavior of the artificial system – whatever it might be – is "observed" over time and analyzed by the researcher.

A major topic of interest is whether certain regularities show up or certain structures evolve on the macro level of the investigated systems. Hence, this leads to the question of whether or not we can observe the evolution of self-organizing behavior of the systems modeled. Self-organization turns out to have two patterns: on the one hand, there are self-regulating processes which are based on negative feed-back and which stabilize given structures; on the other hand, via positive feed-back self-augmenting processes destabilize structures and, by that, might lead to new, innovative structures.[1] The current volume of the series "Lecture Notes in Economics and Mathematical Systems" collects the papers presented in the 9th edition of the Artificial Economics, held in Klagenfurt am Wörthersee (Austria). In particular, this volume contains 18 selected papers. We are very happy that, in addition, one of the keynote speakers, Klaus G. Troitzsch (Universität Koblenz-Landau), found the time summarize his keynote in a paper, which is also included in this volume. The other invited keynote speakers are Andreas Ernst (Universität Kassel) and Gerhard Friedrich (Universität Klagenfurt). All three

---

[1] Witt, Ulrich (1997) Self-organization and Economics – What is New? Structural Change and Economic Dynamics 8(4):489–507. doi:http://dx.doi.org/10.1016/S0954-349X(97)00022-2.

researchers extraordinarily contributed to the development of agent-based models in economics and the social sciences, as well as to its computational foundations.

This volume is divided into six parts. The first part addresses **Methodological Issues**. *Matteo Richiardi* bridges the agent-based modeling approach (of nowadays) with one of its antecedents, i.e. the dynamic microsimulation literature, and elaborates the potential of the latter stream of research for the development in the area of Artificial Economics. In the paper related to his keynote, *Klaus G. Troitzsch* shows – and reminds us of – the importance of (i) testing the significance of simluation results, and (ii) thinking about the variances rather than means of the results.

In the chapter devoted to **Macroeconomics**, first *Hugues Bersini and Nicolas van Zeebroeck* investigate the free market efficiency/equality trade-off by comparing two market mechanisms using an agent-based approach. They find that even though they are more efficient, the competitive (i.e., double auction based) mechanisms tend to increase inequality. *Susanna Calimani and Paolo Pellizzari* model societies where tax evasive behavior of the taxpayers occurs, and analyze the efficiency of different audit policies (of the tax agencies) which depend on the taxpayers characteristics. In their paper *Andrea Teglio, Silvano Cincotti, Einar Jon Erlingsson, Marco Raberto, Hlynur Stefansson, and Jon Thor Sturluson* deal with the current topic of real estate bubbles (as for example observable in the U.S. and in Spain), and investigate the interaction of the level of concentration of financial capital on the formation of real estate bubbles.

The third part of this volume collects four papers related to **Market Dynamics**. *Chih-Hao Lin, Sai-Ping Li, and K.Y. Szeto* investigate an investment strategy based on adaptive trading for anti-correlated pairs of stocks. *Xintong Li, Chao Wang, and Yongui Wang* address the evolution of a decentralized market with network externalities. *Lucian Daniel Stanciu-Viziteu* refines the differentiation between different types of investors (like chartists and fundamentalists). This topic has a remarkable tradition in agent-based models. Stanciu Viziteu distinguishes three types of investors which have different information and make different use of the information. *Wanting Xiong, Han Fu, and Yougui Wang* analyze how fair offers emerge in ultimatum games, and show that fairness considerations as well as adaptive learning are important in the emergence of fair behavior.

In the fourth section of this volume, those papers dealing with **Financial Markets** are comprised. *Olivier Brandouy and Philippe Mathieu* use an agent based model of an artificial stock market in order to analyze the validity of the Volume Synchronized Probability of Informed Trading (VPIN) as a measure for the potential flow toxicity in high frequency markets, and the vicious cycle that might evolve from less informed market makers reacting to flow toxicity. *A. Barazzetti, F. Cecconi, and R. Mastronardi* introduce a predictive machine learning approach based on financial news articles available in the Worldwide Web for event forecasting and trading decisions. In their paper *Michael Roos and Anna Klabunde* introduce findings on the role of trust of angel investors into startup entrepreneurs using an agent-based simulation. Inter alia, they find that neither very high nor very low levels of trust seem to be optimal from the investors' perspective. *Mitja Steinbacher, Matjaz*

Steinbacher, and Matej Steinbacher address a very current topic. They analyze the contagion potential and stability of banking system on a randomized version of the credit contagion model by examining an artificial financial system.

The fifth part of this volume investigates with artificial **Organizations**. *Doris A. Behrens, Silvia Berlinger, and Friederike Wall* employ an agent-based approach in order to analyze how well-known human decision-making biases (e.g., framing effects or the recency effect) in interaction with each other influence overall organizational performance. Utilizing agent-based simulation, *Stephan Leitner and Doris A. Behrens* challenge a well-known economic mechanism for inducing optimal investment decisions, i.e., the competitive hurdle rate mechanism, by testing its robustness in situations where forecasting errors occur. *Marco LiCalzi and Davide Marchiori* shed new light on an "old" but, nonetheless, up-to-date topic: the exploration versus exploitation trade-off. In particular, they revisite recent findings on the multi-armed bandit problem in an environments with high turbulence.

The final section is comprised of three papers related to **Networks**. *Fernando Beltrán and Farhaan Mirza* address the uptake of fibre connections to households and businesses. They model a high-speed, open access broadband network and investigate mutual network effects evolving from the interaction of both end-users and service providers. The contribution of *Sjoukje A. Osinga, Mark R. Kramer, Gert Jan Hofstede and Adrie J. M. Beulens* is a twofold one: On the one hand, the authors investigate the effect of the loss of (market-related) information in a network; on the other hand, the paper provides methodological findings, as it explicitly deals with the advancement of an existing model. Based on mathematical analysis as well as agent-based simulations *Ryota Zamami, Hiroshi Sato, and Akira Namatame* propose a model for designing network structures. In particular, they aim at designing network structures which are robust against systemic risks.

Klagenfurt am Wörthersee, Austria   Stephan Leitner
May 2013   Friederike Wall

Acknowledgements

We would like to thank the following persons who generously donated their time as reviewers and helped in various stages of the production of this volume:

- Simone ALFARANO – Universidad Jaume I de Castellón, Spain
- Frédéric AMBLARD – Université de Toulouse 1, France
- Bruno BEAUFILS – LIFL, Université Lille 1, France
- Doris A. BEHRENS – Alpen-Adria-Universität Klagenfurt, Austria
- Olivier BRANDOUY – University of Paris 1, France
- Dmitri BLÜSCHKE – Alpen-Adria-Universität Klagenfurt, Austria
- Charlotte BRUUN – Aalborg University, Denmark
- Andrea CONSIGLIO – Università degli Studi di Palermo, Italy
- Giorgio FAGIOLO – Scuola Superiore Sant'Anna, Italy
- José Manuel GALÁN – Universidad de Burgos, Spain
- Lynne HAMILL – University of Surrey, United Kingdom
- Florian HAUSER – Universität Innsbruck, Austria
- Cesáreo HERNÁNDEZ – Universidad de Valladolid, Spain
- Jürgen HUBER – Universität Innsbruck, Austria
- Jean Daniel KANT – University of Paris 6, France
- Marco LICALZI – Università "Ca' Foscari" di Venecia, Italy
- Philippe MATHIEU – LIFL, Université Lille 1, France
- Mishael MILAKOVIC – University of Bamberg, Germany
- Sjoukje OSINGA – Wageningen University, The Netherlands
- Juan PAVÓN – Universidad Complutense de Madrid, Spain
- Paolo PELLIZZARI – Università "Ca' Foscari" di Venecia, Italy
- Marta POSADA – Universidad de Valladolid, Spain
- Marco RABERTO – Università di Genova, Italy
- Enrico SCALAS – Università del Piemonte Orientale, Italy
- Klaus SCHREDELSEKER – Universität Innsbruck, Austria
- Andrea TEGLIO – Universidad Jaume I de Castellón, Spain
- Elpida TZAFESTAS – University of Athens, Greece
- Tim VERWAART – Wageningen University, The Netherlands
- Murat YILDIZOGLU – Univ. Montesquieu Bordeaux IV, France

Contents

Part I Methodological Issues

The Missing Link: AB Models and Dynamic Microsimulation 3
Matteo Richiardi

Simulation Experiments and Significance Tests 17
Klaus G. Troitzsch

Part II Macroeconomics

A Stylized Software Model to Explore the Free Market Equality/Efficiency Tradeoff 31
Hugues Bersini and Nicolas van Zeebroeck

Tax Enforcement in an Agent-Based Model with Endogenous Audits 41
Susanna Calimani and Paolo Pellizzari

Subprime Lending and Financial Inequality in an Agent-Based Model ... 55
Andrea Teglio, Silvano Cincotti, Einar Jon Erlingsson,
Marco Raberto, Hlynur Stefansson, and Jon Thor Sturluson

Part III Market Dynamics

Adaptive Trading for Anti-correlated Pairs of Stocks 71
Chih-Hao Lin, Sai-Ping Li, and K.Y. Szeto

Self-Organization of Decentralized Markets with Network Externality ... 81
Xintong Li, Chao Wang, and Yougui Wang

Who Wins: Yoda or Sith? A Proof that Financial Markets Are Seldom Efficient 95
Lucian Daniel Stanciu-Viziteu

Emergence of Fair Offers in Ultimatum Game 107
Wanting Xiong, Han Fu, and Yougui Wang

Part IV Financial Markets

An Agent-Based Investigation of the Probability of Informed Trading 121
Olivier Brandouy and Philippe Mathieu

**Financial Forecasts Based on Analysis of Textual News
Sources: Some Empirical Evidence** 133
A. Barazzetti, F. Cecconi, and R. Mastronardi

Trust in a Network of Investors and Startup Entrepreneurs 147
Michael Roos and Anna Klabunde

**Banks and Their Contagion Potential: How Stable
Is Banking System?** 161
Mitja Steinbacher, Matjaz Steinbacher, and Matej Steinbacher

Part V Organizations

**Phrasing and Timing Information Dissemination
in Organizations: Results of an Agent-Based Simulation** 179
Doris A. Behrens, Silvia Berlinger, and Friederike Wall

**On the Robustness of Coordination Mechanisms for Investment
Decisions Involving 'Incompetent' Agents** 191
Stephan Leitner and Doris A. Behrens

**Pack Light on the Move: Exploitation and Exploration
in a Dynamic Environment** 205
Marco LiCalzi and Davide Marchiori

Part VI Networks

**An Agent-Based Model of Access Uptake on a High-Speed
Broadband Platform** 219
Fernando Beltrán and Farhaan Mirza

**Influence of Losing Multi-dimensional Information
in an Agent-Based Model** 233
Sjoukje A. Osinga, Mark R. Kramer, Gert Jan Hofstede,
and Adrie J.M. Beulens

Least Susceptible Networks to Systemic Risk 245
Ryota Zamami, Hiroshi Sato, and Akira Namatame

Contributors

A. Barazzetti QBT Sagl, Chiasso, Switzerland

Doris A. Behrens Department of Controlling and Strategic Management, Alpen-Adria Universität Klagenfurt, Klagenfurt am Wörthersee, Austria
Department of Mathematical Methods in Economics, Research Unit for Operations Research and Control Systems, Vienna University of Technology, Vienna, Austria

Fernando Beltrán University of Auckland, Auckland, New Zealand

Silvia Berlinger Ilogs AG, Zug, Switzerland

Hugues Bersini IRIDIA-CODE, Université Libre de Bruxelles, Bruxelles, Belgium

Adrie J.M. Beulens Wageningen Univerity, Wageningen, The Netherlands

Olivier Brandouy Sorbonne Graduate Business School, Paris, France

Susanna Calimani Department of Economics and Statistics Cognetti de Martiis, University of Turin, Turin, Italy

F. Cecconi LABSS-ISTC-CNR, Roma, Italy

Silvano Cincotti DIME, Università di Genova, Genova, Italy

Einar Jon Erlingsson Reykjavik University, Reykjavik, Iceland

Han Fu Department of Systems Science, School of Management, Beijing Normal University, Beijing, People's Republic of China

Gert Jan Hofstede Wageningen Univerity, Wageningen, The Netherlands

Anna Klabunde Ruhr University Bochum, Bochum, Germany

Mark R. Kramer Wageningen Univerity, Wageningen, The Netherlands

Stephan Leitner Faculty for Business and Economics, Department of Controlling and Strategic Management, Alpen-Adria Universität Klagenfurt, Klagenfurt, Austria

Sai-Ping Li Institute of Physics, Academia Sinica, Nankang, Taipei, Taiwan, China

Xintong Li Department of Systems Science, School of Management, Beijing Normal University, Beijing, People's Republic of China

Marco LiCalzi Department Management, Università Ca' Foscari Venezia, Venice, Italy

Chih-Hao Lin Institute of Physics, Academia Sinica, Nankang, Taipei, Taiwan, China

Davide Marchiori Department Management, Università Ca' Foscari Venezia, Venice, Italy

R. Mastronardi QBT Sagl, Chiasso, Switzerland

Philippe Mathieu LIFL, UMR CNRS-USTL 8022, Lille, France

Farhaan Mirza University of Auckland, Auckland, New Zealand

Akira Namatame Department of Computer Science, National Defense Academy of Japan, Yokosuka, Japan

Sjoukje A. Osinga Wageningen Univerity, Wageningen, The Netherlands

Paolo Pellizzari Department of Economics, Ca'Foscari University of Venice, Venice, Italy

Marco Raberto DIME, Università di Genova, Genova, Italy

Matteo Richiardi Department of Economics and Statistics, University of Turin, Torino, Italy

Collegio Carlo Alberto and LABORatorio Revelli, Torino, Italy

Michael Roos Ruhr University Bochum, Bochum, Germany

Hiroshi Sato Department of Computer Science, National Defense Academy of Japan, Yokosuka, Japan

Lucian Daniel Stanciu-Viziteu CERAG UMR CNRS 5820, University of Grenoble, Grenoble, France

Hlynur Stefansson Reykjavik University, Reykjavik, Iceland

Matej Steinbacher University of Donja Gorica, Podgorica, Montenegro

Matjaz Steinbacher University of Donja Gorica, Podgorica, Montenegro

Mitja Steinbacher Faculty of Business Studies, Catholic Institute, Ljubljana, Slovenia

Jon Thor Sturluson Reykjavik University, Reykjavik, Iceland

K.Y. Szeto Department of Physics, The Hong Kong University of Science and Technology, Kowloon, Hong Kong SAR, China

Andrea Teglio Departament d'Economia, Universitat Jaume I, Castellón de la Plana, Spain

Klaus G. Troitzsch Institut für Wirtschafts- und Verwaltungsinformatik, Universität Koblenz-Landau, Koblenz, Germany

Friederike Wall Department of Controlling and Strategic Management, Alpen-Adria Universität Klagenfurt, Klagenfurt am Wörthersee, Austria

Chao Wang Department of Systems Science, School of Management, Beijing Normal University, Beijing, People's Republic of China

Yougui Wang Department of Systems Science, School of Management, Beijing Normal University, Beijing, People's Republic of China

Wanting Xiong Department of Systems Science, School of Management, Beijing Normal University, Beijing, People's Republic of China

Ryota Zamami Department of Computer Science, National Defense Academy of Japan, Yokosuka, Japan

Nicolas van Zeebroeck ECARES and IRIDIA-CODE, Université Libre de Bruxelles, Bruxelles, Belgium

Part I
Methodological Issues

The Missing Link: AB Models and Dynamic Microsimulation

Matteo Richiardi

Abstract In this note I pay tribute to two early works by Barbara Bergmann and Gunnar Eliasson which, though firmly grounded in the dynamic microsimulation literature, can be considered as the first examples of large-scale agent-based models. These attempts at building complete micro-to-macro computational models of the economy are important not only in a history of economic thought perspective, but also to encourage convergence of the two approaches in developing credible alternatives to DSGE models.

1 Introduction

Agent-based (AB) models are characterized by three features [24]: (i) there are a multitude of objects that interact with each other and with the environment, (ii) these objects are autonomous, i.e. there is no central, or "top-down" control over their behavior and more generally on the dynamics of the system, and (iii) the outcome of their interaction is numerically computed. AB models are generally identified as *theoretical exercises* aimed at investigating the (unexpected) macro effects arising from the interaction of many individuals—each following possibly simple rules of behavior—or the (unknown) individual routines/strategies underlying some observed macro phenomenon [39]. As such, the typical AB model is a relatively small "toy" model, which can be used to understand relevant mechanisms of social interaction. The roots of AB modeling can be traced down to the study of

---

M. Richiardi (✉)
University of Torino, Department of Economics and Statistics, Campus Luigi Einaudi,
Lungo Dora Siena 100A, 10153 Torino, Italy

Collegio Carlo Alberto and LABORatorio Riccardo Revelli, via Real Collegio 30,
10024 Moncalieri, Torino, Italy
e-mail: matteo.richiardi@unito.it

cellular automata.[1] AB models further developed within the evolutionary economics approach[2] and the so-called Santa Fe perspective on the study of complex systems.[3]

However, an earlier antecedent of AB modelling can be identified in the dynamic microsimulation (DMS) literature, and in particular in two almost forgotten works: Barbara Bergmann's microsimulation of the US economy [6] and Gunnar Eliasson's microsimulation of the Swedish economy [19, 21].

While there has been a recent surge of interest for AB modeling in the DMS literature, this is considered more as a promising direction for future research, than as the continuation of a tradition that dates back to 35 years ago [31, 33]. The infatuation of dynamic microsimulationists for AB modeling is not corresponded by AB practitioners, who not only are not aware of their intellectual debt, but seem not to recognize the convergent paths the two literatures have embraced, with a new vintage of large AB macro-models claiming increasing empirical content.

Both Bergmann and Eliasson developed a macro model with production, investment, and consumption (Eliasson also had a demographic module). They introduced two basic innovations with respect to the DMS literature that was emerging at the time—and in which they were firmly grounded: they explicitly considered the interaction between the supply and demand for labor, and they modeled the behavior of firms and workers in a structural sense. On the other hand, the standard approach to microsimulation—or, as Guy Orcutt called it, the "microanalytic approach for modeling national economies" [36]—was based on the use of what he considered as a-theoretical conditional probability functions, whose change over time, in a recursive framework, describe the evolution of the different processes that were included in the model. This is akin to reduced-form modeling, where each process is analyzed conditional on the past determination of all other processes, including the lagged outcome of the process itself.

Bergmann and Eliasson had a complete and structural, although relatively simple, model of the economy, which were calibrated to replicate many features of the US and Swedish economy, respectively. However, their approach—summarized in [7]—passed relatively unnoticed in the DMS literature, which evolved along the lines identified by Orcutt mainly as reduced form, probabilistic partial equilibrium models, with limited interaction between the micro unit of analysis, and with abundant use of external coordination devices in terms of alignment to exogenously identified control totals. On the contrary, the AB approach emerged with a focus on general equilibrium feedbacks and interaction, at the expenses of richer empirical grounding. Hence, the work of Bergmann and Eliasson could be interpreted as a bridge between the (older) DMS literature and the (newer) AB modeling literature, a bridge that however has so far remained unnoticed. The goal of this paper is to

---

[1]See von Neumann and Burks [45], Gardner [25] and, for a first application to social issues, Schelling [40].
[2]Dosi and Nelson [18].
[3]Anderson et al. [1], Arthur et al. [2], and Blume and Durlauf [8].

bring this bridge back on the map, not only as a tribute to the history of economic thought, but also as a potential useful road for current and future research.

The structure of the paper is as follows. In Sect. 2 DMS is briefly presented; Sects. 3 and 4 are devoted to Bergmann's and Eliasson's models, respectively; Sect. 5 describes how the recent literature on DMS has approached the challenge brought forward by these two precursors; Sect. 6 depicts a convergent evolution of AB models toward increasing complexity and empirical content; finally, Sect. 7 discusses how the emerging approach to estimation of AB models diverges from the one which is dominant in the DMS literature, and suggests that cross-fertilization of techniques might be fruitful.

2 Dynamic Microsimulation

Broadly defined, microsimulation is a methodology used in a large variety of scientific fields to simulate the states and behaviors of different units—individuals, households, firms, etc.—as they evolve in a given environment—a market, a state, an institution. Very often it is motivated by a policy interest, so that narrower definitions are generally provided. For instance, Martini and Trivellato [32] define microsimulation models as "computer programs that simulate aggregate and distributional effects of a policy, by implementing the provisions of the policy on a representative sample of individuals and families, and then summing up the results across individual units" (p. 85).

The field of microsimulation originates from the work of Guy Orcutt in the late 1950s [35]. Orcutt was concerned that macroeconomic models of his time had little to say about the impact of government policy on things like income distribution or poverty, because these models were predicting highly aggregated outputs while lacking sufficiently detailed information of the underlying micro relationships, in terms of the behavior and interaction of the elemental decision-making units. However, if a non-linear relationship exists between an output Y and inputs X, the average value of Y will indeed depend on the whole distribution of X, not on the average value of X only.

Orcutt's revolutionary contribution consisted in his advocacy for a new type of modeling which uses as inputs representative distributions of individuals, households or firms, and puts emphasis on their heterogeneous decision making, as in the real world [37]. In so doing, not only the average value of Y is correctly computed, but its entire distribution can be analyzed. In Orcutt's words, "this new type of model consists of various sorts of interacting units which receive inputs and generate outputs. The outputs of each unit are, in part, functionally related to prior events and, in part, the result of a series of random drawings from discrete probability distributions".

Two things are worth noting. First, the deficiencies of aggregate macro-models identified by Orcutt are still on the table today, more than 50 years later—the recent financial crisis has clarified that the king is naked, exposing all shortcomings of dynamic stochastic general equilibrium (DSGE) models, the workhorse tool in macroeconomics.[4] Second, DMS appears very similar indeed to the AB approach to economic modeling. The main differences can be traced down to the following (i), microsimulations are more policy-oriented, while AB models are more theory-oriented; (ii) microsimulations generally rely on a partial equilibrium approach, while AB models are most often closed models.

As it turns out, in their struggle to replace DSGE models, AB models are becoming more empirically oriented, while microsimulations are becoming more complex, by including more behavioral responses and general equilibrium feedbacks. Bergmann's and Eliasson's models were precursors in the latter respect.

3 Barbara Bergmann's Model of the US Economy

Barbara Bergmann was deeply influenced by Orcutt's lessons while a graduate student at Harvard [34]. However, her microsimulation [6] departs from Orcutt's approach in significant ways. The behavior of all actors is modeled in a structural sense: workers, firms, banks, financial intermediaries, government and the central bank act based on pre-defined decision rules, rather than being described in terms of transition probabilities between different states. Each period (a week), (i) firms make production plans based on past sales and inventory position; (ii) firms attempt to adjust the size of their workforce; wages are set and the government adjusts public employment, (iii) production occurs, (iv) firms adjust prices, (v) firms compute profits, pay taxes and buy inputs for the next period, (vi) workers receive wages, government transfers, property income; they pay taxes and make payments on outstanding loans, (vii) workers decide how much to consume and save, choose among different consumption goods and adjust their portfolios of assets, (viii) firms invest, (ix) the government purchases public procurement from firms, (x) firms make decisions on seeking outside financing, (xi) the government issues public debt, (xii) banks and the financial intermediaries buy or sell private and public bonds; the monetary authority buys or sells government bonds; interest rates are set. In the early 1974 version, only one bank, one financial intermediary and six firms, "representative" of six different types of industrial sectors/consumer goods (motor vehicles, other durables, nondurables, services and construction) were simulated. In the labor market, firms willing to hire make offers to particular workers, some of which are accepted; some vacancies remain unfilled, with the vacancy rate affecting the wage setting mechanism. Unfortunately, the details of the search process are described only in a technical paper that is not easily available anymore [5]. Admittedly, the model was defined by Bergmann herself as a "work in progress", and was completed only years later [4]. The assumption of "representative" firms is

---

[4]See Colander et al. [11], Kirman [29], Krugman [30], Solow [41], and Stiglitz [42].

particularly questionable from an AB perspective, although it is not engraved in the model architecture. However, the model is noteworthy for its complexity and for the ample relevance given to rule-based decision making.

4 Gunnar Eliasson's Model of the Swedish Economy

Eliasson et al. [21] "Micro-to-Macro model", which eventually came to be known as MOSES ("model of the Swedish economy"), is a DMS with firms and workers as the unit of analysis. A concise description of the model can be found in [19]. The labor market module, which is of central importance in the model, is firm-based insofar the search activity is led by the firms that look for the labor force they require to meet their production targets. Labor is homogeneous, and a firm can search the entire market and raid all other firms subject only to the constraint that search takes time (a limited number of search rounds are allowed in each period). Firms scan the market for additional labor randomly, the probability of hitting a source (another firm or the pool of unemployed) being proportional to its size. If a firm meets another firm with a wage level that is sufficiently below its own, it gets the people it wants, up to a maximum proportion of the other firm's labor force. The other firm then adjusts its wage level upwards with a fraction of the difference observed, and it is forced to reconsider its production plan. If a firm raids another firm with a higher wage level it does not get any people, but upgrades its wage offer for the next trial. Firms then produce, sell their products, make investment decisions and revise their expectations. Individuals allocate their income to savings and consumption of durables, non-durables and services. Each year the population is dynamically evolved with flows into and out of the labor force (Fig. 1).

The model was designed to address two issues: (i) formulate a micro explanation for inflation, and (ii) study the relationship between inflation, profits, investment and growth. It was populated partly with real balance sheet firms, and partly with synthetic firms whose balance sheets were calibrated in order to obtain sector totals. Since its original formulation, the model has been updated and documented in a series of papers [20].[5]

---

[5]Of particular relevance here, is the model of the French economy by Gérard Ballot [3]. He models a dual labor market with open-ended and temporary positions. Although the model comprises only 40 firms and 1,700 individuals (belonging to 800 households), it is roughly calibrated to the French labor market over the period 1972–1977, that is around the first oil shock. It is able to reproduce the changes in mobility patterns of some demographic groups when the oil crisis in the 1970s occurred, and in particular the sudden decline of good jobs.

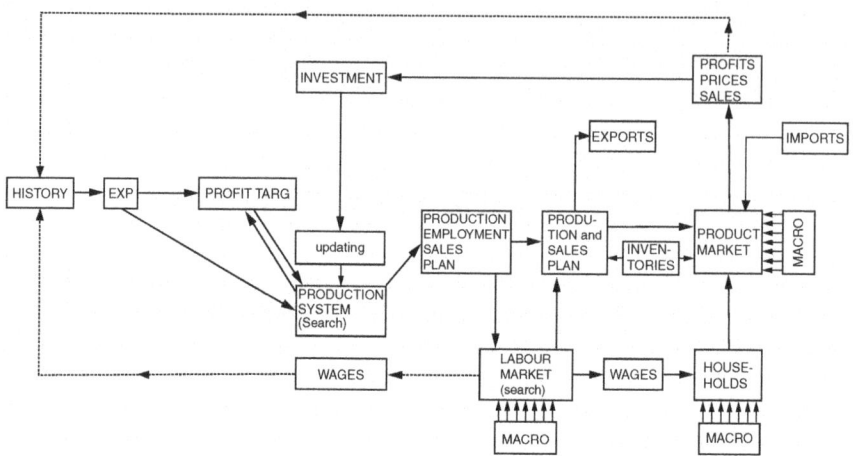

Fig. 1 Structure of the MOSES model of the Swedish economy (Source: Eliasson [19])

5 Current Trends in DMS Modeling: Linking Micro to Macro

Rather than following the strategy of increasing the complexity of microsimulation models and explicitly model general equilibrium feedbacks—along the lines pioneered by Bergmann and Eliasson—the recent literature on DMS has tried to link partial equilibrium microsimulations with computable general equilibrium (CGE) macro models. The underlying idea is to keep the models as simple as possible, and develop different models for different levels of aggregation. Following again Orcutt's insights, these models could then be connected through intermediate variables. Peichl [38] describes how the approach, labeled *top-down bottom-up*, works: the CGE model produces macroeconomic variables (price level, output growth rates, etc.) which are passed as inputs to the DMS; the microsimulation model in turns produces outcomes (elasticities, income, etc.) which are passed back to the CGE; the procedure is repeated until convergence (Fig. 2).

Simple and appealing as the approach may look, it is plagued by theoretical and empirical inconsistencies, which might preclude convergence or, worse, produce outcomes which are misleading for policy analysis (the Lucas critique once again). The approach is also computationally burdensome, and only few applications have so far been developed.[6]

---

[6]See Peichl [38] for a review.

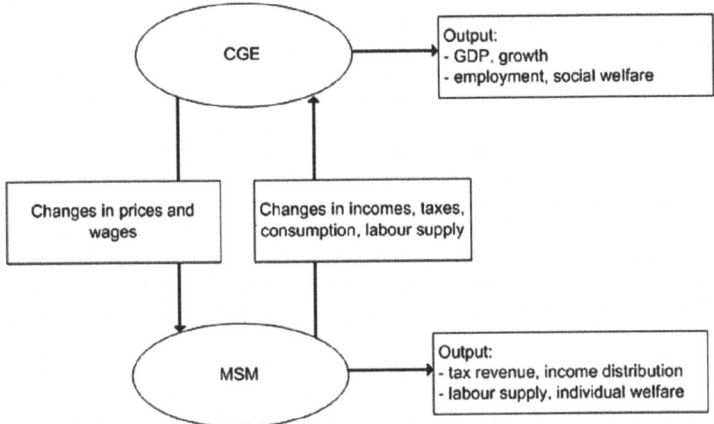

Fig. 2 Linking microsimulation models (MSM) and CGE models: the top-down bottom-up approach (Source: Peichl [38])

6 Current Trends in AB Macro Modeling: Towards More General Models

Recent years have witnessed a trend in AB macro modeling towards more detailed and richer models, targeting to a higher number of stylized macro facts, and claiming a stronger empirical content.[7] A big push forward has been provided, on the East side of the Atlantic, by two large projects funded by the European Commission: EURACE and CRISIS.[8] The three year (2006–2009) EURACE project had the ambitious goal of creating an integrated AB model of the European economy [17], linking real (consumption goods, investment goods and labor) and financial (loans, bonds and stocks) markets. The model was meant to be populated by a very large number of fairly sophisticated agents (in the order of 10^7 households, 10^5 consumption goods producers, 10^2 investment goods producers, and 10^2 banks), each following empirically documented behavioral rules. This very large number of agents allows in principle to discover emerging phenomena and/or rare events that would not occur with a smaller population.

---

[7] See Dawid et al. [16] for a review.

[8] The CRISIS project is still in its initial phase; its goal is to provide "a platform for the development and application of data mining, process mining, computational and artificial intelligence and every other computer and complex science technique coupled with economic theory and econometric methods that can be devoted to identifying the emergence of social and economic risks, instabilities and crises" [22].

An important feature of the model is its explicit spatial structure: with the exception of the investment goods market and the asset market, all markets are local. For example, there is a local labor market in each NUTS-2 region. Deissenberg et al. [17] succinctly describe how the model works:

> The market for consumption goods is a decentralized market, with local interaction between the firms and consumers. We assume that the firms send their merchandise to a given set of local shopping malls. All buying and selling occurs at these malls. Firms chose the outlet malls on the basis of expected local demand and profit opportunities. They also take into account the costs involved in servicing a particular mall, such as the transportation costs, the leases for the stores in the mall, and the inventory management costs.
>
> The labor market is also a decentralized market. A local search-and-matching process is used to represent the interaction between firms and workers. The firms post vacancies, including the minimum skill level required for the posted job. The potential employees apply to vacancies that have been posted by firms in their local neighborhood. Unemployed workers who do not succeed in finding a job locally can migrate to a different region.
>
> The market for investment goods is a centralized market. There are multiple investment goods producers, each producing a different, vertically differentiated, technology. The investment goods producers invest in R&D to technologically improve the investment goods, leading to oligopolistic competition among them. The producers of consumption goods can invest in one of these technologies to produce a variety of differentiated consumption goods.
>
> On the credit market, the firms interact with banks to obtain loans. The credit market is a decentralized market, with competition between banks setting different interest rates for the business loans. The banks apply credit standards to the firms that apply for the loans. Thus, the firms can be credit constrained.
>
> Finally, the financial asset market links the real side with the financial side. Firms issue equity (common stocks and corporate bonds) to finance investments and production. The households invest in asset portfolios, and the government sells government bonds to finance its budget deficit. The financial market thus consists of a market for corporate and government bonds and a market for firm stocks. The linkage between the financial side and the real side of the economy is provided by the financial policy of the firms on internal and external financing, that is among others, by the dividend, the debt repayment, and the investment decisions.

Figure 3 shows the interactions between producers and consumers in the markets for investment and consumption goods.

Admittedly, EURACE reached a level of complexity rarely seen in an economic model, and proved difficult to manage. As the EU funding run out, the project developed into smaller scale models, each maintained by a different research unit. Herbert Dawid and his team, at the University of Bielefeld, focused on skills formation and innovation; their model was upgraded [12, 14] and gave rise to a steady stream of publications.[9] Silvano Cingotti and his team, at the University of Genoa, focused on the credit market and bank regulation.[10] The team lead by Mauro

---

[9] See Dawid et al. [13, 15].

[10] Cincotti and Teglio [10] and Teglio et al. [43].

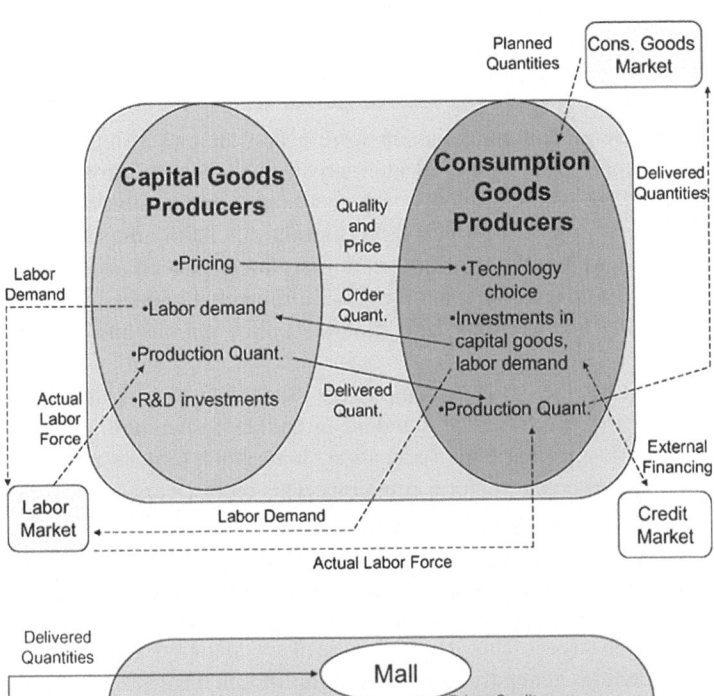
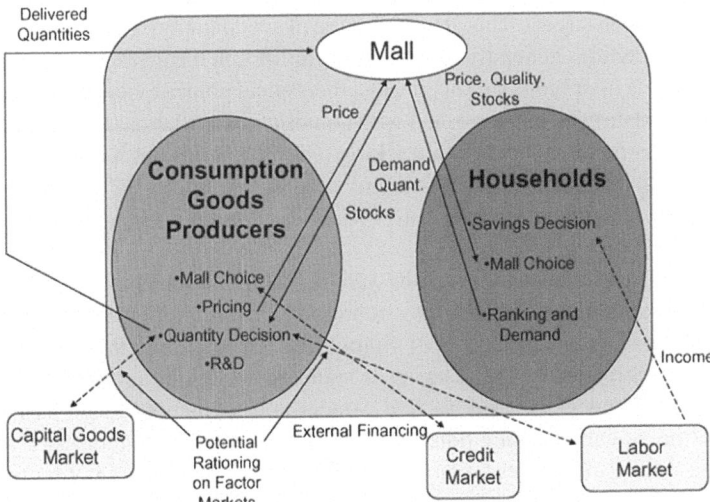

Fig. 3 Interactions in the capital goods market (*top panel*) and the consumption goods market (*bottom panel*) (Source: Deissenberg et al. [17])

Gallegati and Domenico Delli Gatti, at the Marche Polytechnic University and the Catholic University, adopted another strategy and kept on developing ad-hoc models for specific applications.

7 Microsimulation vs. AB Modeling: The Challenge of Empirical Validation

The works by Bergmann and Eliasson were a first attempt at replacing the core of macroeconomics [9] with an AB alternative. Their goal to provide a structural closed model of the whole economy, to be calibrated empirically, was indeed very ambitious. After more than 30 years, the literature is taking the challenge again. The models reviewed in the previous section are more general and more complex than their overlooked ancestors, but they will ultimately be judged under the same metric: the ability to track real data, simulate policy options and provide policy guidance.

With respect to the issue of empirical validation, two approaches can be identified. The first one, which is standard in the DMS literature, is to separately estimate the different processes (education, household formation, labor market participation, etc.) as reduced-form equations. This requires to assume that choices are made sequentially, so that all the covariates in every process can be considered as predetermined. This assumption being often untenable, the practical solution is to keep the estimates separate and "adjust" the estimates (via *alignment* algorithms) in order to keep the evolution of some macro-variables of interest in line with exogenously given targets. This also takes care of specification errors, and of the fact that microsimulations generally lack general equilibrium feedback effects. However, the procedure is of dubious validity, from a theoretical point of view. It is considered as a "quick and dirty" solution to deal with complicated models and inadequate data; it can (by construction) succeed in tracking real time series but it has no structural backbone and thus it is likely to fail to predict the effects of policy changes; moreover, it is not able to provide out-of-sample guidance when no external targets are available.

The second approach is the one followed by DSGE modeling, which has evolved from rough calibration to structural estimation [23, 44]. Notwithstanding the fact that DSGE models are packed with simplifying assumptions, their estimation is by no means straightforward. The most standard technique is ML and requires to linearize the model in order to find a local approximation of the steady state solution (the so-called *policy functions*), then express this solution as a Markov chain (the *state space representation*), then apply filtering theory in order to obtain the likelihood function. Because this likelihood function is generally very flat and quite uninformative about the underlying structural parameters, more curvature is added by introducing Bayesian priors.

As for AB models, empirical validation is still rare, and generally amounts to more or less sophisticated calibration; only a few applications exist where a structural estimation of an AB model is performed, and they normally involve very simple models.[11] The use of simulated minimum distance estimators appear to be a

---

[11] See Grazzini et al. [27, 28] and the references therein.

promising approach [26], but the feasibility of the approach in a large scale model has still to be proved.

This test will eventually decide the future of macroeconomics. If the structural estimation of large scale AB models remains beyond reach, a sequential approach might become dominant, where—following the standard practice in the microsimulation literature—different submodules are separately estimated. This will however dent the appeal of AB models as an alternative to DSGE models in Economics. If, on the other hand, large scale AB models prove amenable to structural estimation, it is likely that they will eventually encompass DMS and establish their validity—over and beyond DSGE models—not only to explore theoretical possibilities but also to analyze policy relevant issues.

Acknowledgements Financial support from Collegio Carlo Alberto is gratefully acknowledged.

References

1. Anderson P, Arrow K, Pines D (eds) (1988) The economy as an evolving complex system. SFI studies in the sciences of complexity. Addison-Wesley Longman, Redwood City
2. Arthur W, Durlauf S, Lane D (eds) (1997) The economy as an evolving complex system II. Addison-Wesley Longman, Reading
3. Ballot G (2002) Modeling the labor market as an evolving institution: model artemis. J Econ Behav Organ 49:51–77
4. Bennett RL, Bergmann BR (1986) A microsimulated transactions model of the United States economy. John Hopkins University Press, Baltimore
5. Bergmann BR (1973) Labor turnover, segmentation and rates of unemployment: a simulation-theoretic approach. Technical report, University of Maryland
6. Bergmann BR (1974) A microsimulation of the macroeconomy with explicitely represented money flows. Ann Econ Soc Meas 3(3):475–489
7. Bergmann BR, Eliasson G, Orcutt GH (eds) (1977) Micro simulation–models, methods, and applications: proceedings of the symposium on micro simulation methods, Stockholm, 19–22 Sept
8. Blume L, Durlauf S (eds) (2006) The economy as an evolving complex system, III. Current perspectives and future directions. Oxford University Press, Oxford
9. Caballero RJ (2010) Macroeconomics after the crisis: time to deal with the pretense-of-knowledge syndrome. J Econ Perspect 24:85–102
10. Cincotti, Silvano RM, Teglio A (2010) Credit money and macroeconomic instability in the agent-based model and simulator eurace. Economics: Open-Access, Open-Assess E-J 4(2010-26). http://www.economics-ejournal.org/economics/journalarticles/2010-26
11. Colander D, Howitt P, Kirman A, Leijonhufvud A, Mehrling P (2008) Toward an empirically based macroeconomics. Am Econ Rev Papers Proc 98(2):236–240
12. Dawid H, Neugart M (2011) Agent-based models for economic policy design. East Econ J 37(1):44–50
13. Dawid H, Gemkow S, Harting P, van der Hoog S, Neugart M (2009) On the effects of skill upgrading in the presence of spatial labor market frictions: an agent-based analysis of spatial policy design. J Artif Soc Soc Simul 12:4
14. Dawid H, Gemkow S, Harting P, van der Hoog S, Neugart M (2011) The eurace@unibi model: an agent-based macroeconomic model for economic policy design. Technical report, University of Bielefeld

15. Dawid H, Gemkow S, Harting P, Neugart M (2012) Labor market integration policies and the convergence of regions: the role of skills and technology diffusion. J Evol Econ 22(3):543–562
16. Dawid H, Gemkow S, Harting P, van der Hoog S, Neugart M (2013) Agent-based macroeconomic modeling and policy analysis: the eurace@unibi model. In: Chen S-H, Kaboudan M (eds) Handbook on computational economics and finance. Oxford University Press, Oxford (forthcoming)
17. Deissenberg C, van der Hoog S, Dawid H (2008) Eurace: a massively parallel agent-based model of the European economy. Appl Math Comput 204:541–552
18. Dosi G, Nelson R (1994) An introduction to evolutionary theories in economics. J Evol Econ 4:153–172
19. Eliasson G (1977) Competition and market processes in a simulation model of the Swedish economy. Am Econ Rev 67:277–281
20. Eliasson G (1991) Modeling the experimentally organized economy: complex dynamics in an empirical micro-macro model of endogenous economic growth. J Econ Behav Organ 16:153–182
21. Eliasson G, Olavi G, Heiman M (1976) A micro-macro interactive simulation model of the swedish economy. FörvaltningsbolagetSindex
22. Farmer DJ, Gallegati M, Hommes C, Kirman A, Ormerod P, Cincotti S, Sanchez A, Helbing D (2012) A complex systems approach to constructing better models for managing financial markets and the economy. J Eur Phys J: Spec Top 214(1):295–324
23. Fernández-Villaverde J (2010) The econometrics of dsge models. SERIEs 1:3–49
24. Gallegati M, Richiardi M (2009) Agent-based modelling in economics and complexity. In: Meyer B (ed) Encyclopedia of complexity and system science. Springer, New York, pp 200–224
25. Gardner M (1970) Mathematical games: the fantastic combinations of john conwa's new solitaire game 'life'. Sci Am 223(4):120–123
26. Grazzini J, Richiardi M (2013) Consistent estimation of agent-based models by simulated minimum distance. Working Paper 130/2013, LABORatorio R. Revelli
27. Grazzini J, Richiardi M, Sella L (2012a) Small sample bias in msm estimation of agent-based models. In Teglio A, Alfarano S, Camacho-Cuena E, Ginés-Vilar M (eds) Managing market complexity. The approach of artificial economics. Lecture notes in economics and mathematical systems. Springer, Heidelberg
28. Grazzini J, Richiardi MG, Sella L (2012b) Indirect estimation of agent-based models. An application to a simple diffusion model. Complex Econ 1(2): (forthcoming)
29. Kirman A (2010) The economic crisis is a crisis for economic theory. CESifo Econ Stud 56(4):498–535
30. Krugman P (2011) The profession and the crisis. East Econ J 37:303–312
31. Li J (2011) Dynamic microsimulation for public policy analysis. Boekenplan, Maastricht
32. Martini A, Trivellato U (1997) The role of survey data in microsimulation models for social policy analysis. Labour 11(1):83–112
33. Morand E, Toulemon L, Pennec S, Baggio R, Billari F (2010) Demographic modelling: the state of the art. SustainCity Working Paper 2.1a, Ined, Paris
34. Olson PI (2007) On the contributions of Barbara Bergmann to economics. Rev Pol Econ 19:475–496
35. Orcutt GH (1957) A new type of socio-economic system. Rev Econ Stat 39:116–123
36. Orcutt GH (1990) The microanalytic approach for modeling national economies. J Econ Behav Organ 14:29–41
37. Orcutt GH, Greenberger M, Korbel J, Rivlin AM (1961) Microanalysis of socioeconomic systems. A simulation study. Harper & Brothers, New York
38. Peichl A (2009) The benefits and problems of linking micro and macro models. Evidence from a flat tax analysis. J Appl Econ XII(2):301–329
39. Richiardi M (2012) Agent-based computational economics: a short introduction. Knowl Eng Rev 27(2):137–149
40. Schelling T (1969) Models of segregation. Am Econ Rev 59(2):488–493

41. Solow R (2010) Building a science of economics for the real world. Subcommittee on investigations and oversight, committee on science and technology, hearing of Jul 20, 2010 10:00am to 12:00pm
42. Stiglitz JE (2011) Rethinking macroeconomics: what failed, and how to repair it. J Eur Econ Assoc 9:591–645
43. Teglio A, Raberto M, Cincotti S (2012) The impact of banks' capital adequacy regulation on the economic system: an agent-based approach. Adv Complex Syst 15(suppl02):361–400
44. Tovar CE (2008) Dsge models and central banks. BIS Working Papers 258, Bank for International Settlements
45. von Neumann J, Burks A (1966) Theory of self-reproducing automata. University of Illinois Press, Urbana

Simulation Experiments and Significance Tests

Klaus G. Troitzsch

Abstract The paper uses two formal models of simple processes in artificial worlds—one with and one without an analytical solution—to discuss the role of statistical analysis and significance tests of results from multiple simulation runs. Moreover the paper argues that it is not the sheer existence of an effect of input parameters on simulation results but the effect *size* which is the interesting outcome of a simulation and that significance tests of differences in means are much less important than the distribution of output variables which are more often than not non-normal distributions.

1 Introduction

Simulation models, and particularly agent-based simulation models, often come in the form of stochastic models where at least micro-entities perform their actions not deterministically but only with a certain probability which is often derived from expected utilities of the actions available at the time a decision is made. This leads to the fact that simulation results do not only depend on fixed meaningful parameters but also on the seed of a random number generator, such that results of single simulation runs are often classified as insufficient and a larger number of simulation runs is deemed necessary to explore the simulation model.

When relations between input parameters of such stochastic simulation models and simulation outcomes are analysed the question occurs whether such a relation is "statistically significant", and more often than not one wonders how many simulation runs are necessary for a valid analysis of the simulation outcome. Obviously, both questions cannot be answered at the same time. Increasing the

---

K.G. Troitzsch (✉)
Institut für Wirtschafts- und Verwaltungsinformatik, Universität Koblenz-Landau,
Koblenz, Germany
e-mail: kgt@uni-koblenz.de

number of parallel simulation runs (i.e. runs with the same combination of input parameters but with different random number generator seeds) automatically leads to a numerical improvement of the "level of significance" without leading to an improved meaningfulness: a correlation or regression coefficient of 0.1 may be significantly different from 0.0 on a certain level when the number of simulation runs whose results enter into the analysis is above a certain threshold (which, of course, depends on the distribution of the coefficient and the chosen level of significance). Thus any statement about the relation between input parameters and simulation output can be "made significant" no matter what the effect size is, as there is always a fixed relation between effect size, significance level and sample size (here: number of simulation runs).[1]

This paper discusses the role of simulation "experiments" in contrast to laboratory and field experiments and the role of a "sample" of simulation runs in contrast to samples drawn from real populations or universes. We will come to the conclusion that multiple runs of otherwise identical simulations which only differ in the random seed (or other technical or physical means of random number generators) are samples in the mathematical meaning of the term (and in so far superior to samples in the social and economic sciences drawn from real populations which are most often biased by low response rates, self-selection and other obstacles to perfect sampling) but that significance testing is not an adequate means of analysing simulation results. This is partly due to the fact that simulation is not an inductive method but a method of deducting new statements from the "first principles" (theoretical assumptions) incorporated into the simulation model, more or less with the same purpose as deriving a formula of a time dependent probability density function—a macro property—from assumptions about the stochastic micro behaviour. This idea leads to a further result: the average of a coefficient or parameter over a large number of runs of the same simulation model is less interesting than its distribution as a whole, and comparing averages with each other or with zero is sometimes useless.

The paper will use two very different models to illustrate the roles of effect sizes and distribution functions as results of simulation models and wherever possible compare these roles to the role of mathematical analysis. But unfortunately this is often only possible in simple models whereas in economics and the social sciences one is more often interested in more complex models. The first of these two models describes a lock-in process mainly with the methods of mathematical analysis and dates back to the 1970s when Wolfgang Weidlich and Hermann Haken applied methods of statistical physics to social and economic phenomena. This model has meanwhile been superseded by more sophisticated models which do not even have

---

[1]For an in-depth discussion of "the cult of statistical significance" see [8]. Particularly, Ziliak and McCloskey criticise the use of sentences like "The differences reached the level of statistical significance, by and large." [p. 35]. Their general argument is that it is not enough to "decide '*whether* there *exists* an effect' " [p. 25], but that it is necessary to ask "the scientific question 'How much is the effect?' ".

an approximate analytical solution. The second model (of which several different aspects will be discussed) is a model of the role of minimum wages in low-cost sectors of the economy. The comparison of the two models will show the role of simulation as a deductive method.

2 The Role of Analytically Solvable Models

A simulation model can be seen as a formalisation of a theory, and in terms of the 'non-statement view' it can also be seen as a *full model of a theory* which in turn was specified according to the rules laid down in [2] (see also [4]). Thus simulation is a method of deduction, much like the classical mathematical deduction, for cases when analytical deduction does not lead to a closed solution. As most agent-based models (and, indeed, most simulation models) have stochastical components, one can define the deduction of an estimate of a probability or probability density function as one of the main tasks of a simulation. In relatively simple multilevel models this deduction can be done analytically as the following example (originally provided by Weidlich and Haag [7]) shows which deduces the time dependent probability density function for the proportions of proponents and opponents of a certain decision from a microspecification [3] of the behaviour of the micro-entities which can choose between 'yes' and 'no' with a probability (Eq. 1) depending on the current majority (Eq. 3).

$$\mu_{yes \leftarrow no} = \nu \exp(\delta + \kappa x)$$
$$\mu_{no \leftarrow yes} = \nu \exp[-(\delta + \kappa x)] \quad (1)$$

$$n = n_{yes} - n_{no} \quad (2)$$

$$x = \frac{n_{yes} - n_{no}}{n_{yes} + n_{no}} \quad (3)$$

From this primary assumption one can derive the probabilities w that the whole population changes its state (measured as $n = n_{yes} - n_{no}$ where traditionally $n_{yes} + n_{no} = 2N$).

$$w[(n+1) \leftarrow n] = w_\uparrow(n) = n_- \mu_{yes \leftarrow no} = (N-n)\mu_{yes \leftarrow no} \quad (4)$$
$$w[(n-1) \leftarrow n] = w_\downarrow(n) = n_+ \mu_{no \leftarrow yes} = (N+n)\mu_{no \leftarrow yes} \quad (5)$$
$$w[j \leftarrow i] = 0 \quad \text{for} \quad |i-j| > 1 \quad (6)$$

Combining the probabilities p of the population to be in a certain state n with the probabilities w of changing its state leads to the master equation determining the probability of the population of being in one of its possible states $n \in \{-N, \ldots, N\}$ at time t:

$$\frac{p(n;t+\Delta t) - p(n;t)}{\Delta t} = p(n+1;t)w_\downarrow(n+1)$$
$$-p(n;t)(w_\uparrow(n) + w_\downarrow(n))$$
$$+p(n-1;t)w_\uparrow(n-1) \qquad (7)$$

which, by taking the limit $\Delta t \to 0$ and further simplification, yields the system of linear differential equations[2] for $2N+1$ functions $p(n;t)$:

$$\dot{\mathbf{p}}(t) = \mathbf{L}\mathbf{p}(t) \qquad (8)$$

where $\mathbf{p}(t)$ is a vector of the probabilities $p(n;t)$ for all the possible population states, and $\mathbf{L}$ is a matrix which has non-vanishing elements only in the main diagonal and in the two adjacent diagonals, and all its elements are constant:

$$\begin{aligned}
l_{ii} &= -w_\downarrow(i) - w_\uparrow(i) \\
l_{ij} &= w_\downarrow(j) & j &= i+1 \\
l_{ij} &= w_\uparrow(j) & j &= i-1 \\
l_{ij} &= 0 & |i-j| &> 1
\end{aligned}$$

This, by the way, leads to $\sum_i \dot{p}(i;t) = 0$ for all t, which also fulfils the condition $\sum_i p(i;t) = 1$.

Equation 8 is a system of coupled linear differential equations and could be solved by analytic means, although it is solved numerically here because for a population size of $2N$ the system consists of $2N+1$ equations. By analytic means, however, the stable equilibrium distribution of populations for $t \to \infty$ may be calculated approximately, where the approximation is fairly good for population sizes above 50, and its minima and maxima can be depicted as a function of the two parameters δ and κ, see Fig. 2.

κ represents the strength of the coupling of the individuals to the majority and determines whether a population is likely to have a 50-50 distribution of 'yes' and 'no' ($\kappa < 1$ for $\delta = 0$) or is likely to have a strong majority of either 'yes' and 'no' ($\kappa > 1$ for $\delta = 0$). With $\delta \neq 0$ and small κ, the most probable majority in a population would be different from 50 % (and with high κ the probability maxima in the right-hand part of Fig. 1 would be of different height). ν is a frequency parameter but it is of little interest: it affects only the time scale of the structure-building process, because with higher ν the breakthrough of either 'yes' or 'no' comes faster. For $\delta = 0$ and $\kappa > 1$ the distribution of populations develops into a bimodal

---

[2] A numerical solution of this system of differential equations is a simulation of the macro object 'population' with the vector-valued attribute 'probability of being in one of the possible states'.

Fig. 1 Time-dependent probability density functions of the Weidlich-Haag process for two different κ's

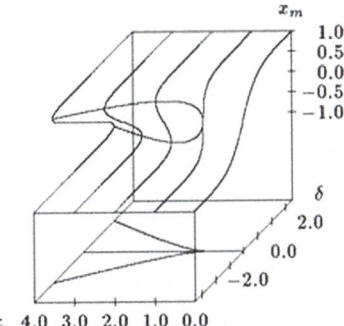

Fig. 2 Dependence of the stationary probability density function of the Weidlich-Haag process for varying δ and κs. x_s is the set of maxima and minima of the stationary probability density function

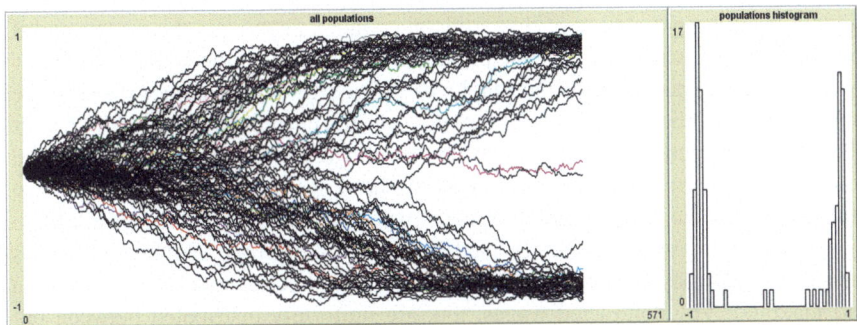

Fig. 3 Trajectories of 100 populations with 200 members each from a simulation of the Weidlich-Haag process for $\kappa = 1.5$s and the histogram of the x distribution

distribution—the probability of finding the population with a strong majority of either 'yes' or 'no' is very high. For $\kappa < 1$ the probability of an evenly split population is very high (see Fig. 1—for $\delta \neq 0$ the threshold for κ is different). Figure 2 shows the dependence of the minima and maxima of the solutions to Eq. 8 on the two parameters κ and δ.

Whereas a stochastic simulation of the process can visualise the time-dependent probability distribution shown in the right-hand plot of Fig. 1—see Fig. 3 which shows a high degree of similarity—an analogue to Fig. 2 is much more difficult to generate from just simulation.

After this short digression into the world of analytically solvable models we can return to strictly formalised models whose solution cannot be found with standard analytical means of what Ostrom [6] called the second symbol system—namely that of mathematics, the first being natural language—but which call for methods taken from the third symbol system, the one of programming languages and computer simulation. Apart from the fact that Fig. 1 had already been output of a numerical simulation (namely the iterative evaluation of Eq. 7) the central difference is now that the numerical values of the $p(n; t)$ can no longer be determined; instead we run a number of realisations of the stochastic process defined in the simulation model and try to reconstruct the shape of the probability or probability density functions from this 'sample' of simulation runs.

3 The Role of Simulation Models

The second model has the same purpose as the first. It deduces the effect of one or two parameters controlling micro behaviour on the outcome which in this case, too, is the probability distribution of one or two macro variables.[3] The model describes the behaviour of employers seeking workers in three different low-cost sectors and of workers seeking employment in one or another of these sectors. In every tick, one employer agent and one worker agent (currently employed or not), both selected at random from the population of this artificial economy, negotiate the wages, and if they find that the wages is above the minimum expectation of the worker and below the maximum expectation of the employer they agree on a contract (in case the worker agent had been employed before it leaves the current employment in favour of the higher wages offered by the new employer agent). Both employer agents and worker agents form their expectations on the base of information they have from other agents in the market such that their expectations range from the minimum to the maximum wages currently paid by or to a selection of their colleagues (and the wages they bid or offer are uniformly distributed between these minima and maxima). The contracted wages, too, are uniformly distributed between bid and offer. Thus a stochastic influence comes into play in three different phases of each simulation step. One external parameter of the model is the minimum wages which serves as a lower bound to both sides' expectations.[4] In the results reported in the next few paragraphs, contracted wages can be as small as the minimum wages which are valid in this artificial economy (even wages near 0 can be contracted, but given that in modern societies something like a basic income is paid to everybody,

---

[3]This model goes back to discussions between Gregor van der Beek, two master students and the author, was partly documented in the two students' master thesis [5] and extended by the present author.

[4]The current version of the model can be found among the NetLogo User Community Models, http://ccl.northwestern.edu/netlogo/models/community/MinimumWages.

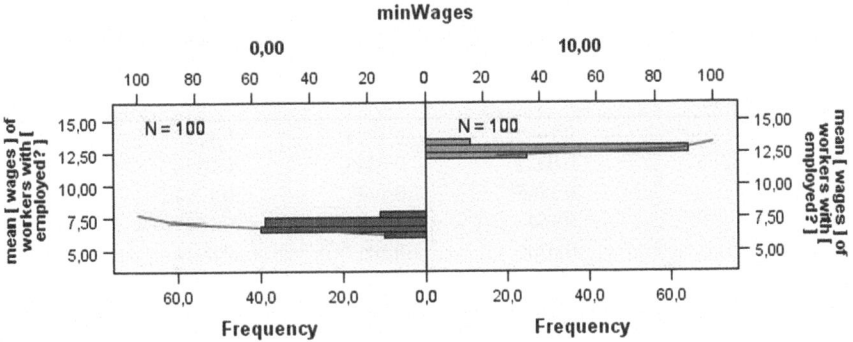

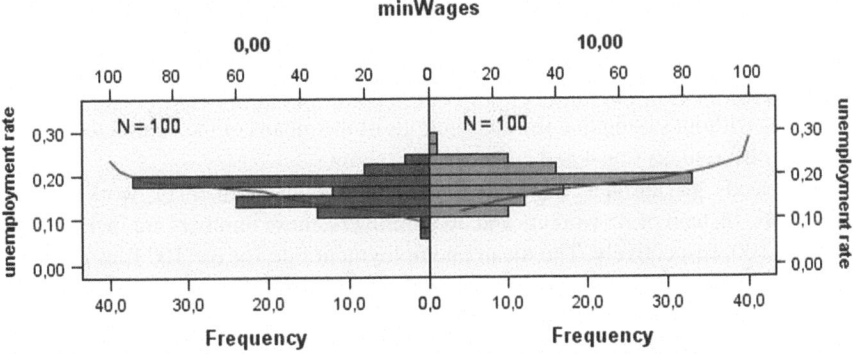

Fig. 4 The effect of minimum wages on average wages and unemployment rate

this makes the following simulation runs slightly unrealistic, for an extension see below).

A first experiment with this model shows the difference between 100 runs each without minimum wages and with a minimum wages of 10 (think of per hour)—see Fig. 4. At first glance one is not surprised to find that the average wages are higher with a minimum wages of 10, but one is a little doubtful whether the minimum wages had an effect on the unemployment rate. The Kolmogorov-Smirnov test says that the null hypothesis that the two distributions are the same should be rejected with an α of 0.037, and a t-test says the same for the comparison of the two means with $\alpha < 0.0005$. Note that the two α's are very different! The left-hand side of Table 1 shows some more information on the two distributions—one should add that the distribution for no minimum wages has two modes for 0.1529 (0.15) and 0.1765 (0.16) with a minimum in between at 0.1647 (0.12). Thus one could believe that the distribution in the no minimum wages case is bimodal, such that the mean is as irrelevant as it is in the case of the lock-in model for higher

Table 1 The effect of minimum wages on the unemployment rate in two different scenarios

	Workers seek ...			
	Random employer Minimum wages		Best employer Minimum wages	
Unemployment rate	0	10	0	10
Mean	0.1667	0.1844	0.1518	0.1521
Standard deviation	0.0327	0.0348	0.0320	0.0298
Mode	0.1765	0.1647	0.1294	0.1412
Median	0.1647	0.1765	0.1529	0.1529
Minimum	0.0706	0.1059	0.0941	0.0824
Maximum	0.2471	0.2824	0.2235	0.2353
Range	0.1765	0.1765	0.1294	0.1529

κ—but with only 85 workers in the current example the two modes with 13 and 15 unemployed and the minimum with 14 unemployed are not convincing. And, moreover, the Kolmogorov-Smirnov test says that the null hypothesis that in both cases the distribution is a normal distribution should not be rejected ($\alpha = 0.300$ and 0.218, respectively).

It goes without saying that the distributions of the means of the contracted wages are very different, as Fig. 4 and later on Fig. 6 show.

This leads to another experiment with a higher number of workers and employers. Instead of 85 workers and 30 employers, these numbers are increased to 300 and 100, respectively. The mean unemployment rate for the 100 runs each for the two values of minimum wages is 0.098667 and 0.149933, respectively, with a standard error of about 0.017 in both cases.

Thus it seems that the unemployment rate is normally distributed over the parallel runs of the simulation with a larger artificial economy, and mean and (perhaps) standard deviation are different for different values of the parameter minimum wages. The question remains whether this statistically significant difference is meaningful at all. The statistical significance estimated from our 100 runs for the two values of the parameter minimum wages means that the hypothesis derived from the simulation model that there is no effect of minimum wages on the unemployment rate has to be rejected at an $\alpha < 0.0005$ (according to the Kolmogorov-Smirnov two-sample test whose test statistics is 6.081 in this case).

To answer this question we return to the original size of our artificial economy (where the difference between the two unemployment rates was considerably smaller (see left part of Table 1) and check the unemployment rates for more values of the parameter minimum wages—besides 0 and 10, also 2, 4, 6 and 8. Figure 5 shows visually that the effect of the minimum wages on the unemployment rate is all but overwhelming.

Analysing this result further, we find that the regression equation between the minimum wages w and the unemployment rate u is $u = 0.162 + 0.002w$, with a standardised regression coefficient of 0.213 and a reduction of variance (R^2) of 0.046 (both of which, too, are significantly different from 0, $\alpha = 0.000195$). Simply formulated this means that a percentage point increase of 1 in the minimum wages

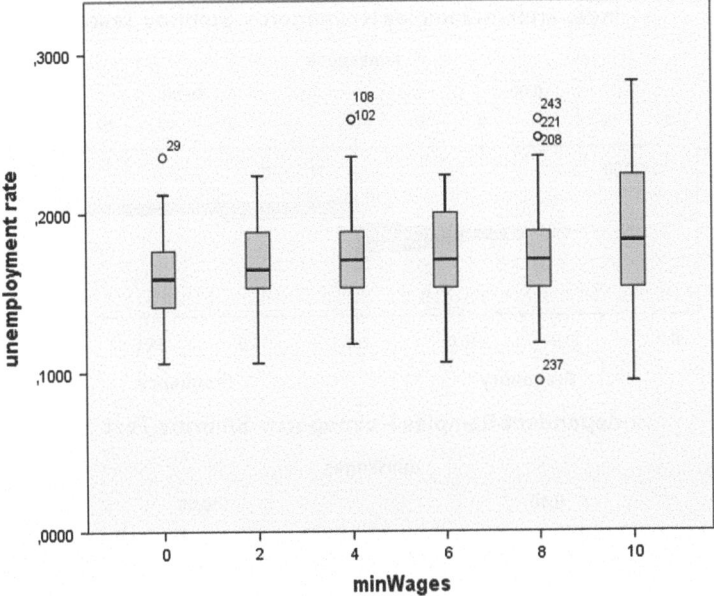

Fig. 5 The effect of different values of the parameter minimum wages on the unemployment rate

leads to an increase of the unemployment rate of 0.2 % points. 95.4 % of the variance of the unemployment rates in the $6 \times 50 = 300$ runs remains unexplained by the minimum wages parameter and can only be explained by the random noise in the model.

The conclusion from this first model setting is: There is a statistically significant effect, but it is small.

Now let us change the setting a little. In the runs discussed so far workers and employers met randomly and started their negotiations, and both only knew the range of wages contracted in the past among their colleagues, formed their ask and bid wages by a random selection of a number within the respective range, and when the employer offered more than the worker had expected, the contract was executed. In what follows we will be a little more realistic: The worker agent randomly selected in every time step knows beforehand how much the individual employers offer and selects the employer with the highest offer (which could perhaps be less than the worker's expectation, but in this case no negotiation is started). Again the contractual wages will be a random value, uniformly distributed between the wages expected by the worker and the offer of the employer—which is the best currently offered. One stochastic element is removed, as compared to the former version, namely the random assignment between workers and employers. This leads to the following result (for an easier comparison documented in the right-hand side of Table 1).

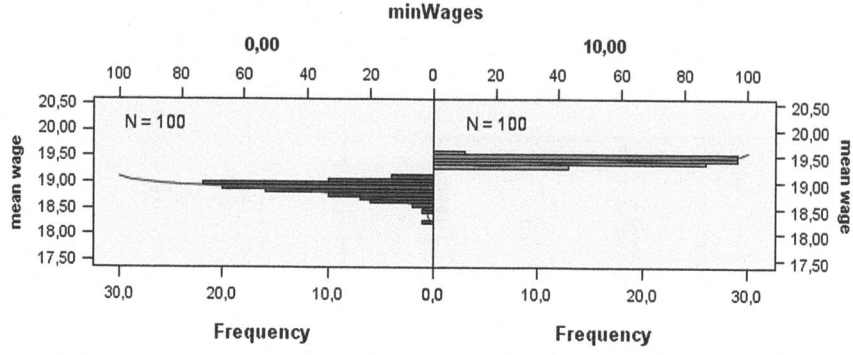

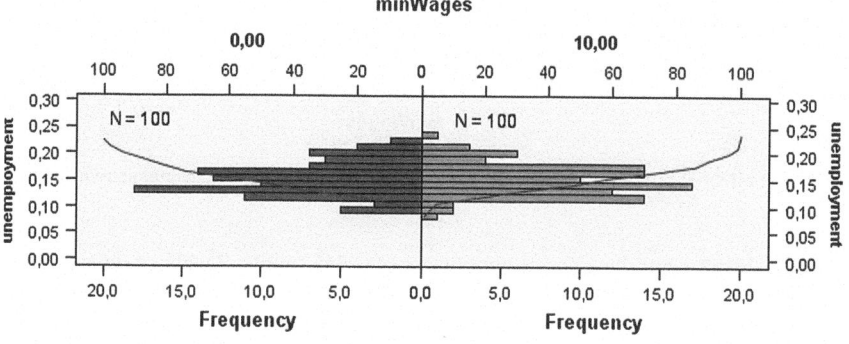

Fig. 6 The effect of minimum wages on average wages and unemployment rate

The median unemployment rate is exactly the same in this setting both with and without minimum wages, and the two distributions also seem to be the same, as the Kolmogorov-Smirnov test suggests to retain this hypothesis ($\alpha = 0.994$, test statistic 0.424, see also Fig. 6 which also shows that the distributions of the average wages for the two values of the minimum wage parameter are different). The hypothesis that the distributions of the unemployment rates are normal again cannot be rejected ($\alpha = 0.77$ and 0.96, respectively), but the histograms are not very similar to a histogram of a normally distributed variable.

Thus under slightly more realistic conditions, the simulation reveals that the introduction of minimum wages does not influence the unemployment rate. This does, of course, not exclude that with a higher number of simulation runs or with a larger size of the employer and worker populations the situation changes. As with the current results the probability distributions over the parallel simulation runs seems to be a normal distribution—which does not come as a surprise, as the random effects included in the simulation model are more or less of the same type and only linearly related—one can expect that the statistical significance, measured as α of the small differences between means will change and that the difference of 0.000353 between the two means will become "statistically significant" when the number of runs is increased to 2,000 instead of 100 for each parameter value. But the question remains whether it is reasonable to run the model twenty times as often to make an unemployment rate percentage point difference of 0.0353 "significant" on the $\alpha = 0.05$ level.

4 Conclusion

Our examples showed that a stochastic simulation model can be used to find out what the distribution of result variables is like and to determine its parameters.

In the case of the first example (where an approximate analytical solution is available in parallel to a multitude of simulation runs) the simulation showed that under certain parameter combinations a bimodal distribution emerges from the assumptions of the model, but it would perhaps be difficult to find out for which κ threshold the bifurcation into a bimodal distribution happens. And it is perhaps impossible to claim from simulations that this threshold is $\kappa = 1.0$—perhaps the best guess from simulation runs with 500 steps as in Fig. 3 is $0.7 < \kappa < 1.3$. This shows that wherever an analytical solution is possible simulations should only be used for illustrative visualisations. Thus one can still support Alker's verdict that simulation is inelegant mathematics [1]. Moreover one cannot decide how many simulation runs are necessary to find out that for $\kappa > 1$ a bimodal distribution emerges—one run is certainly not enough but would only yield the information that the population will quickly show a strong majority of one of the two alternatives.

In the case of the second example where not even an approximate analytical solution is available to the stochastic process and where the standard macro method of finding an equilibrium at the intersection of the demand and supply curves makes one believe that the introduction of minimum wages leads to a higher unemployment rate, we see that the simulation supports the macro derivation in so far as with actual wages between 0 and some maximum and no directed search for an employer agent which offers the highest wages, the unemployment rate is slightly less than in the more realistic version where workers do not contract randomly but only with the employer who currently offers the highest wages. Here the median was identical between the simulation runs with and without minimum wages, and all other comparisons showed no "significant" differences. The distribution over 100 runs for each parameter set shows that two single simulation runs are by no means sufficient to evaluate the difference. In a way the situation in this case is the same as the comparison between two regions of the same country in one of which minimum wages were introduced while in the other region this was not the case. The two regions would certainly show two different rates of unemployment, but given that unemployment rates are influenced by so many other environmental variables this simple "field experiment" comparison should not convince anybody. The multitude of simulation runs, however, allows for a comparison not of two single realisations of two stochastic processes but of two distributions of two stochastic processes.

But what does "significant" mean in this context? For the third comparison of the second example we expect that in 1 out of 2,000 simulation runs (for both values of the minimum wages) we would have found a difference in the mean which a standard t-test would have identified as significantly different from 0. The probability that by pure chance the one and only simulation run had this extraordinary outcome, and the probability that we had restricted ourselves to this singular simulation run is even much less (as one simulation run on a standard PC takes less than 2 min). Thus if we compare significance considerations between analyses of empirical data (where we usually have only one potentially biased sample) and analyses of simulation output (where we can arbitrarily increase the number of usually unbiased samples) we see that significance is not significant in the latter case.

References

1. Alker HR Jr (1974) Computer simulations: inelegant mathematics and worse social science? Int J Math Educ Sci Technol 5:139–155
2. Balzer W, Moulines CU, Sneed JD (1987) An architectonic for science. The structuralist program. Volume 186 of synthese library. Reidel, Dordrecht
3. Epstein JM, Axtell R (1996) Growing artificial societies – social science from the bottom up. MIT, Cambridge MA
4. Ihrig M, Troitzsch KG (2013) An extended research framework for the simulation era. In: Diaz R, Longo F (eds) 2013 Spring simulation multiconference on emerging M&S applications in industry and academia symposium and the modeling and humanities (EAIA and MathH 2013) (SpringSim '13). Simulation Series, vol 45(5). Curran Associates, Red Hook, pp 99–106
5. Jazayeri P, Tohum MH (2012) Auswirkung der Einführung eines Mindestlohns auf den Arbeitsmarkt, anhand eines Simulationsmodells. http://kola.opus.hbz-nrw.de/volltexte/2012/759/
6. Ostrom T (1988) Computer simulation: the third symbol system. J Exp Soc Psychol 24:381–392
7. Weidlich W, Haag G (1983) Concepts and models of a quantitative sociology. The dynamics of interacting populations. Springer series in synergetics, vol 14. Springer, Berlin
8. Ziliak ST, McCloskey DN (2007) The cult of statistical significance. The University of Michigan Press, Ann Arbor

Part II
Macroeconomics

A Stylized Software Model to Explore the Free Market Equality/Efficiency Tradeoff

Hugues Bersini and Nicolas van Zeebroeck

Abstract This paper provides an agent-based software exploration of the well-known free market efficiency/equality trade-off. Our study simulates the interaction of agents producing, trading and consuming goods in the presence of different market structures, and looks at how efficient the producers/consumers mapping turn out to be as well as the resulting distribution of welfare among agents at the end of an arbitrarily large number of iterations. Two market mechanisms are compared: the competitive market (a double auction market in which agents outbid each other in order to buy and sell products) and the random one (in which products are allocated randomly). Our results confirm that the superior efficiency of the competitive market (an effective and never stopping producers/consumers mapping and a superior aggregative welfare) comes at a very high price in terms of inequality (above all when severe budget constraints are in play).

1 Introduction

A classical disputed question regarding the effect of free market economy on the social welfare is the right balance between equality and efficiency called by Okun [7]: the big tradeoff. *"If both equality and efficiency are valued, and neither takes absolute priority over the other, then, in places where they conflict, compromises ought to be struck. In such cases, some equality will be sacrificed for the sake of*

H. Bersini (✉)
IRIDIA-CODE, Université Libre de Bruxelles, CP 194/6, 50 av. Franklin Roosevelt,
1050 Bruxelles, Belgium
e-mail: bersini@ulb.ac.be

N. van Zeebroeck
ECARES and IRIDIA-CODE, Université Libre de Bruxelles, CP 114/04,
50 av. Franklin Roosevelt, 1050 Bruxelles, Belgium
e-mail: nivzeebr@ulb.ac.be

efficiency and some efficiency for the sake of equality. But any sacrifice of either has to be justified as a necessary means of obtaining more of the other". Okun introduced the metaphor of the leaky bucket now famous among economists, which explains how favouring equality by money transfer is very likely to decrease the size of the pie to be shared.

Part of the problem lies in the difficulty to appropriately define these two notions. The eternal question of equality, famously debated and popularized by, among the most modern thinkers, Rawls, Dworkin [4], Sen [8], depends upon (1) the right currency for equality (primary goods, consumers utility, opportunity, achievements, capability, ...) and (2) the right distribution of this currency (pure equality (for instance the Nash's maximization of the utility products [2]), some form of minmax principles i.e. favoring at a given time a distribution that is to the greatest benefit of the least-advantaged agent or others).

On the other hand, the question of economic efficiency is even more ambiguous. It was originally framed around the Pareto optimality for which no one well-being should be raised without as a consequence reducing someone else well-being.

Now many Pareto optima can be obtained on an imaginary axis, going from a pure utilitarian aggregative end (at which what really counts is to maximize the collective (i.e. aggregative) well-being) to a more equalitarian end (where what really counts is to maximize the well-being of the worst agent or their utility product). Indeed Pareto optimum per se is totally unconcerned with the appropriate distribution of the economical profit. It is enough that all agents welfare simply grow in time as a result of the economical interactions, leaving completely unresolved the comparison of economic systems that either promote aggregate welfare, perfect equality or the improvement of the poorer to the expense of the richer.

Another quite classical definition of efficiency related with multi-agents competitive system is the "allocative" one, in which the system must guarantee that a resource is being produced by the most skillful producer (at the smallest cost) and goes to someone who draws the greater utility out of it. Not surprisingly, although efficient according to this definition, such a competitive system, likely to promote the best producers and to feed the greediest consumers, may have very little chance to equally distribute wealth.

Beyond this historical debate about which economical system (free or regulated) has to be privileged between an "aggregative" or a "distributive" one, there is another key efficiency criteria which is often left out of the discussion, originally due to Hayek's pioneering insights: his metaphor of the market as a "system of telecommunication" [3]. Market prices are primarily a means of collating and conveying information for the producers to adequately response to the consumers needs that should be left uncorrupted by state intervention. Thus, though a very high price prevents most of the consumers to acquire a product, it is, in the same time, a very reliable source of information addressed to the producer that many consumers are desperately in need of such a product. It might well be possible that a distributive and more regulated economy, flattening the prices and rendering most of the products affordable to almost all, and although morally very defendable, turns out to corrupt this efficient self-organised distributed information transmission mechanism and makes all economical agents to see their situation finally degrade in time on account of a misguided production. In the rest of the paper, we will designate such incapacity of our simulated markets to effectively map producers onto consumers as market failures (MF). Those will raise as runtime exceptions of our software model.

In order to address these different issues, an object oriented software stylized model is proposed comparing two very different structures of market that potentially should drive the collective welfare to the two extremes: aggregative on one side and distributive on the other. These two structures are first a double auction competitive market (in which buyers and sellers compete to outbid each other) and a random market (in which the matching between buyers and sellers as well as their transaction is done in a purely random way). Following the description of the model, many experimental outcomes of a large amount of robust runs (giving quite constant results) will be presented along three key dimensions: the Gini indices (regarding equality), the aggregate utility and the probability of market failures (the last two more concerned with efficiency).

2 The Model

As illustrated in the UML class diagram of Fig. 1, the model maps onto a C# object oriented software where the distinct responsibilities have been distributed through the many classes. Although UML has become the de facto standard for the graphical visualization of software development and has seen its usage constantly rising during all these years, still the UML diagrams remain very absent of most of the economics publications [1]. It is regrettable since UML proposes a set of well defined and standardized diagrams (transcending any specific programming language and any computer platform) to naturally describe and resolve problems on the basis of the high level concepts inherent to the formulation of the problem. It is enough to discover and draw the main actors ("word", "agents", "behaviors", "resources", "physical or conceptual site") of the problem and how they do mutually relate and interact in time to seriously progress towards the algorithmic solution of this problem.

The main encompassing class, the World, contains one Market, either competitive or random, where a given number of agents have the opportunity to successively produce, sell, buy and consume. This world evolves through discrete ticks. At every tick, a randomly selected agent is given a chance to produce one unit of one product among n possible ones. The market then attempts to execute one transaction that involves one buyer and one seller marketing one unit of a given product. If no transaction turns out to be possible after a rather huge number of attempts, on account of an impossible pairing between buyers and sellers, the model raises a market failure (equivalent to a runtime exception in the C# program). Once acquired by the buyer, the product is immediately consumed during the same tick and converted into utility according to his associated taste. This is the only way

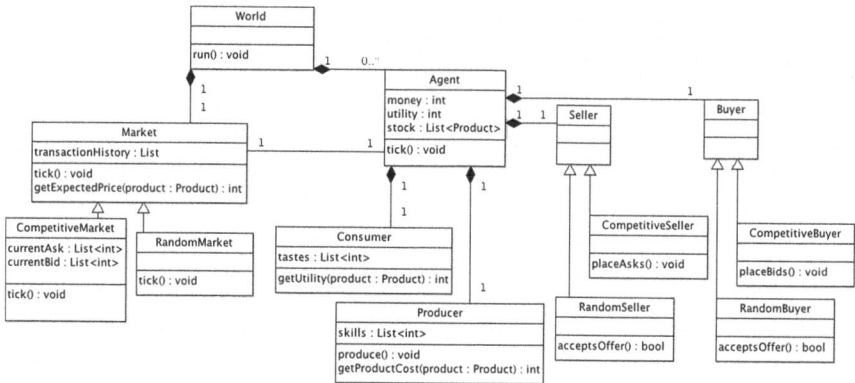

Fig. 1 The UML class diagram of the software model

for the agent's utility to grow up. Every agent starts with the same amount of money at the beginning of the simulation (allowing him to produce goods). Agents are distinctively characterized by two crucial factors randomly distributed, which are their skills (influencing their producing behavior, production prices amount to the skills) and their tastes (imprinting their consuming behavior, utility increase amount to the tastes). While individual skills and tastes, taken randomly between 0 and 1, vary among agent, the initial total amount of skills and tastes are normalized to 1. This is the departing point of agents differentiation during the simulation and the only initial cause for any further inequality growing among the agents. However, it is quite important to insist that all agents are initially set as equal in utility and richness and just slightly different in the way they like and elaborate products. Both producer and consumer behaviors are strictly similar in the competitive and the random markets whereas seller and buyer behaviors become fundamentally different.

Once randomly selected, the producer first has to decide which product to make. Two factors influence his decision: his skills and the average price of the last m transactions. Knowing his skills to produce each product unit and the average price in the market (memorized during the m previous ticks), it is obvious to compute his expected profit for each product. After x productions of the same product, an agent can further specialize himself making the production cost randomly diminishing within a moving range. Skills are then renormalized to one with all other skills proportionally rising up. Once an agent buys a product, it is immediately consumed, with effect to increase the agent utility by the value of his taste for this product.

The competitive market is akin to a continuous double auction market [5, 6], in which agents bid to buy and sell products units. During a succession of steps, the market repeatedly invites two randomly selected agents to place asks and bids on one product they want to sell or purchase. At the first tick, the market is initialized with best-buying and best-selling offers for all the products on the market (bids at price null and asks at price max). Then a random seller is selected to place an ask for the most profitable product he has in stock (the proposed price should be below the best-selling offer and incurring the least expense (i.e. selecting the product with the highest skill), this price is finally set between the producer skill and the current best-selling offer). The market then looks whether this ask crosses the current best-buying offer on that particular product. If so, the transaction occurs, if not, the ask becomes the best-selling offer and the market turns to the buying part. The randomly selected competitive buyer shows the very symmetrical behavior. He first selects the most desirable product (one with the highest taste above the best-buying offer) and places a bid limited by his reservation price (the proposed price is set between the best-buying offer and the reservation price). The market looks whether this bid crosses the current best-selling offer. Once two offers cross, the transaction price is fixed as the buying offer price. If following a determined number of trials, no transaction is to be found, a market failure is reported. The random market is much simpler, since the sellers and the buyers behave without particular interest. In this version, a random seller places an ask on a random product, on which a random buyer is invited to react. If the buyer reservation price is higher than the price asked by the seller, a transaction takes place, the price being randomly set between the two offers. Here again, if following a determined number of trials, no transaction turns out to be possible, a market failure is reported.

Finally, in order to impose a budgetary constraint on the buyers behavior, the reservation price for any product is fixed as the taste multiplied by the current money endowed by the agent multiplied by a time index (the agent portion of the budget he wills to engage at every tick). Of course, in all cases, bids and asks are only posted if the agent has, respectively, enough money to cover it or has a unit of the product in stock (as a result of previous productions). Whatever initial conditions being set: number of agents, number of products, vector of tastes and skills for every agent, initial endowment of money for all agents, they are obviously exactly equal for both market simulations, the objective being to compare the competitive version of the market (supposedly more efficient) with the random one (supposedly more equalitarian) only on the basis of the market structure and the agent's behaviours.

3 The Results

Four key metrics can be measured out of the different simulations: utility (increasing by consumption), money (leaking out by production and then fluctuating according to the transactions), added value (the difference between the price earned by the seller and the production cost) and market failures. For the first three, the aggregate value over all agents is used as an indicator of the market efficiency while the Gini coefficient (computed again for all three) testifies of how unequal this market turns out to be. The market failures (labeled MF in the following) is also used as an indicator of the market efficiency, but in the sense originally given by Hayek.

Our simulations are always executed in the presence of 50 agents, 10 products and during 50,000 simulation steps.

For the first set of simulations, each agent is endowed with 500 units of money (so no budgetary constraint is imposed at all) and the number of past transactions kept in memory to inform the producer on the most valuable products is 1,000. Additionally the consumers do not see their taste decreasing in time as an outcome of their consumption. Typical and quite robust experimental results follow, first for the random market then the competitive one.

Random Market: Total Utility: 5,390, Total Money: 24,312, Gini Utility: 0.04, Gini Money: 0.007, MF: 0

Competitive Market: Total Utility: 9,755, Total Money: 24,491, Gini Utility: 0.27, Gini Money: 0.08, MF:0

The competitive market turns out to be much more efficient in aggregative terms but this superior efficiency comes at a very high price in terms of inequality, compared with a random market (the utility Gini index is seven times greater as a result of the competition). Distortions in utility and money tend to grow over time. The competitive market favors those with skill in demand and those with taste skillfully expressed. Efficiency makes gifted agents much more likely to produce and greedy ones much more likely to consume. Given that this difference in taste can be continuously expressed over the simulation, a self-amplifying pairing happens between the greedy consumers and their "dedicate" competent producers.

In the case of a marginally decreasing consuming utility, results become quite different, now making the competitive and the random markets rather comparable.

Random Market: Total Utility: 5,152, Total Money: 24,244, Gini Utility: 0.02, Gini Money: 0.007, MF: 0

Competitive Market: Total Utility: 5,424, Total Money: 24,488, Gini Utility: 0.042, Gini Money: 0.004, MF:0

In the absence of any budgetary constraint and if the same tastes cannot be differentially expressed all over the simulation (since being alternatively up and down as a result of the consumption), the competitive and random markets turn out to be very equivalent both in terms of efficiency and equality. One first, somewhat unexpected, conclusion of our simulations is that in the presence of a decreasing marginal utility and in the absence of any budgetary constraint (agents remain sufficiently rich during the whole duration), no real outcome difference is to be pointed out between a competitive and a random market. For the remaining of the simulations and in agreement with classical economics, we will maintain a decreasing utility, making the agents seing their taste adjusted in time as an outcome of their consumption.

The next aspect that deserves a dedicate treatment is the impact of information on the competitive market, evaluated by gradually varying the number of past transactions taken into account during the production process (fixed to 1,000 so far) i.e. the quality and the reliability of the information available to the producers to guide their productions towards the real consumers needs. Many simulations have been run where the producers exploit an increasing number of past transactions: 0, 1, 5, 10, 50, 100, 500, 1,000, 5,000, 10,000, 50,000. In the previous simulations

A Stylized Software Model to Explore the Free Market Equality/Efficiency Tradeoff

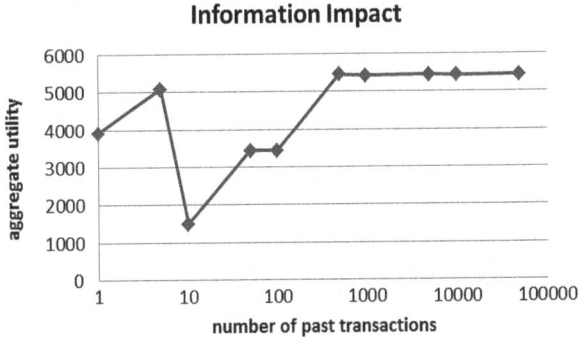

Fig. 2 Effect of the number of past transactions taken into account to optimize the production on the aggregate utility of the market

discussed so far, this number has been settled to 1,000. We compute the average aggregate utility as a function of this number and, surprisingly, as shown in Fig. 2, the resulting curve is not monotonous. Below 100 past transactions available to the producers, and quite robustly, the resulting competitive markets show an important number of failures with a pick at 10, demonstrating, unexpectedly, that a total ignorance of the past is even better than a very little knowledge. An increasing amount of information first dilutes the effective signal upon which producers base their decisions. Producers in those cases may be better off only focusing on their own costs than on their expected profits. We finally can observe the relevance of sufficient information for the competitive market to efficiently allocate the available resources (1,000 past transactions seem to be an appropriate minimal threshold above which no improvement is observed).

The last and most relevant aspect of the model to be explored is the influence of the budgetary constraint on the behavior of the market. While maintaining all other features constant (50 agents, 10 products, information based on 1,000 past transactions), the initial money endowment is being decreased: 100, 80, 60, 50, 25, 20, 15, 10. After showing many difficulties in running until the end of the simulation, the random version of the market simply stops executing at around an initial endowment of 25. Many agents go bankrupt and the simulation is being constantly interrupted by market failures. Although very egalitarian, both facts once again testify of the inefficiency of the random market to map the producers onto the consumers. The producers waste their money making products that the consumers appear definitely not interested in.

As regards the competitive version, the table below in Fig. 3 indicates how the budget constraint really impacts the model as the initial endowment decreases. Although an initial budget of 20, 15 or 10, entails a few intermittent market failures, the model can now always keep running over the 50,000 simulation ticks. The most striking fact of this table is the evolution of the utility Gini index as well as the added value one (for instance they respectively reach a pick of 0.25 and 0.20 for

Money	100	80	60	50	25	20	15	10
Utility	5465	5340	5464	5452	5433	5401	5330	5246
Money	4182	3275	2093	1735	494	284	83	17
Ad. Val.	3680	3720	3264	3380	2675	2400	1817	1424
G(Util)	0.05	0.061	0.077	0.070	0.088	0.12	0.16	0.25
G(Mon)	0.006	0.007	0.013	0.010	0.016	0.017	0.014	0.08
G(AV)	0.08	0.092	0.11	0.10	0.13	0.15	0.17	0.20
MF	0	0	0	0	0	26	270	914

Fig. 3 Summary of results (aggregate and Gini) obtained by gradually decreasing the initial budget possessed by every agent

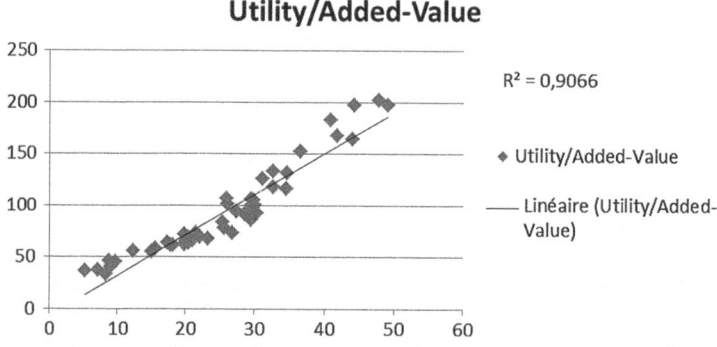

Fig. 4 Correlation between the added value of the producer and his final utility as a consumer (established over all 50 agents)

an initial budget of 10 by agent) that clearly shows a growing inequality as the money becomes scarcer. Again the market keeps being efficient but now to the large expense of equality. The competitive regime becomes much more selective towards the most skillful producers, the only ones who are effectively able to compete in the market. Budget constraint and money scarcity decrease the potential gains for producers but above all redirect them towards the best producers. Moreover, specialization acts as given the best producers even more marketing power. Budget constraints make the competition so severe that any tiny difference in skills is identified and reinforced. Figure 4 interestingly shows the correlation between the added value of the producer and his final utility as a consumer (i.e. established over all 50 agents). A clear positive correlation is observable between the added value of the producer and his consumption (90 % of the utility distribution is explained by the added value distribution). The greediest and most satisfied consumers turn out to be the best producers. As observable in Fig. 5, showing the evolution of the utility Gini index, inequality among the agents is on a fast growing trend. Competitive market acts in self-amplifying the market dominance of producers who can benefit from even the tiniest initial comparative advantage.

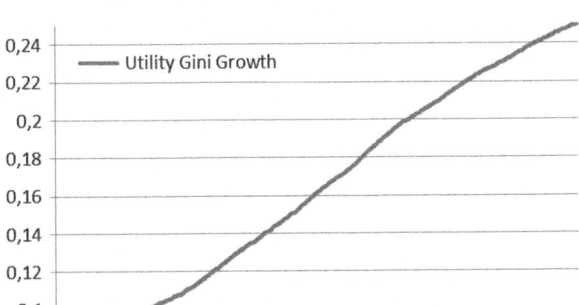

Fig. 5 Evolution in time of the Utility Gini Index for an initial budget of 10

4 Conclusions

This paper describes a stylized simulation exercise in which we compare a double auction, quite aggressive, competitive market, with a pure theoretical abstraction that represents a market in which producer and consumer matching is purely made on a random basis (under a natural set of constraints: budget constraint, no sale at loss rule for the producers). Our main simulation results confirm the higher efficiency generally attributed to competitive markets first to simply map the consumers onto the producers then in maximizing the aggregative welfare. However in most of the studies of competitive markets, very little attention is paid to the equality in welfare distribution (even if such inequality is well known to become explosive these recent years, accompanying the international spreading of free market economy). Our results equally show this inequality explosion, above all in the case of budgetary constraints, when only the best producers can survive, make money and consume. Interestingly enough, at the starting of our simulation, all agents can be considered as equally ready and gifted to take part in the market, but its inherent competitive structure (in contrast with the random one) makes an even negligible difference in skills to be greatly amplified with time.

In line with most of the ethical philosophers, we can easily argue about the immoral nature of such an inequality amplifier mechanism (even when equality of opportunity is fully guaranteed) and the definitive need for a complementary equalizing system. In most of our modern societies, such systems are generally assumed by a central organization benefiting from the profit of the winning agents to redistribute to the unlucky ones. However, our results equally show that any source of randomness can also (and perhaps more easily, since still in the absence of any centralization and deliberation) compensate for the inequality self-reinforcing tendency coming along with competition. Such irrational intrusions are already at work to soften the effect of free market since competing agents have limited time and cognitive resources to explore all possible offers and many apparently unexplained motivations undermine a lot of trading decisions, to the benefit of all.

References

1. Bersini H (2012) UML for ABM. J Artif Soc Soc Simul 15(1):9
2. Binmore K (2005) Natural justice. Oxford University Press, New York
3. Cassidy J (2009) How markets fail. Penguin Books, London, England
4. Dworkin R (1981) What is equality? Part 1: Equality of welfare. Philos Public Aff 10(3): 185–246
5. Fanoi S, Pellizzari P (2011) Time-dependent trading strategies in a continuous double auction. In: Osinga S, Hofstede GJ, Verwaart T (eds) Emergent results of atificial economics. Springer, Berlin/Heidelberg, pp 165–176
6. Gode DK, Sunder S (1993) Allocative efficiency of markets with zero-intelligence traders: market as a partial substitute for individual rationality. J Political Eco 101(1):119–137
7. Okun AK (1975) Equality and efficiency, the big tradeoff. The Brookings Institution, Washington, DC
8. Sen AK (1973) On economic inequality. Oxford University Press, USA

Tax Enforcement in an Agent-Based Model with Endogenous Audits

Susanna Calimani and Paolo Pellizzari

Abstract We generalize the classic Allingham and Sandmo's model of tax evasion considering heterogeneous agents with different degrees of tax morale and matchable, as opposed to non-matchable, income. The Tax Agency evolves its control scheme, maximizing the revenues from fines, and takes into account some minimal information on the taxpayers. We compare different audit policies and find that the most effective scheme remarkably depends on the way agents update the subjective probability of being audited, on the distribution of matchable income in the population as well as on the level of tax morale. Hence, different features of societies and taxpayers' behaviors not only affect the compliance rate, as expected, but require the Tax Agency to alter its audit policy in a context-dependent way. In particular, high revenues are obtained performing random audits when agents think they are directed towards peculiar individuals and, conversely, should be biased towards low declarations when taxpayers believe audits are nonspecific or random.

1 Introduction

Tax evasion has always been a dear issue to policy makers, but in times of crisis when Governments fall short of resources, it easily becomes one of the favorite pieces in everybody's political programme. It is clear that tax evasion dynamics should be investigated taking into account the joint action of many heterogeneous taxpayers and of the Government (or, equivalently, of the agency that collects taxes).

S. Calimani (✉)
Department of Economics and Statistics "Cognetti de Martiis", University of Turin, Turin, Italy
e-mail: susanna.calimani@unito.it

P. Pellizzari
Department of Economics, Ca' Foscari University of Venice, Venice, Italy
e-mail: paolop@unive.it

However, starting with the seminal paper [1], taxpayers' behavior has been at the centre of the focus, using models with fully rational agents who decide how much of their actual income to declare to the tax authority, in an expected utility framework à la von Neumann and Morgenstern. The conduct of the Government was instead simply mimicked using a constant fine rate and assuming that entirely random auditing is performed with some probability. This basic model well captures the deterrent effects of increased audit probability and fines, but typically predicts much higher evasion than it is observed and suggests that hikes in the tax rate would increase compliance, whereas intuition would lead to the opposite outcome. An impressive number of alternative models tried to overcome these shortcomings, resorting, say, to the tendency of *overweighting small probabilities*, [7], introducing a psychological cost of evasion ("shame") in the utility function and developing more general paradigms in which the *tax morale* of a community plays a role, see [10]. The interaction between agents and Government can also be framed in terms of power and trust, see [8], where the former increases the *enforced* compliance and the latter can boost *voluntary* compliance.

In this paper, we exploit the flexibility and heterogeneity of an agent-based approach to show that the inclusion of more realistic features notably improves the descriptive accuracy of the standard model. The agent-based model enables us to depict a system of many heterogeneous agents, whose interaction might give rise to emergent phenomena that are not analytically derivable, in particular when the model exhibits potential network effects, nonlinear behavior, learning or adaptation. These features characterize a multitude of social phenomena, and certainly also the tax evasion framework we examine: in such a contest, we use agent-based simulations to grasp the effects that different combinations of societal configurations and audit policies might have.

First, we allow the tax agency to evolve its own control scheme, using minimal information about taxpayers' income to improve the efficacy of the audits. The weight of these pieces of information can, in fact, be calibrated to optimize who is to be inspected.

Second, our agents are restricted in their evasion decisions and, in particular, we assume that some fraction of income cannot be concealed. Income can be categorized as traceable or non-traceable: the first includes salary or wage as well as self-employed income that can be matched by somebody else's tax-report, whereas the second is formed by all those earnings' components that are hardly matchable and easily hidden. Therefore, we emphasize in this work the distinction of matchable versus non-matchable income, as done in [5] where it is empirically shown that the non-matchable component notably increased in recent years for US taxpayers, with visible repercussions on aggregate compliance. Take, for instance, two taxpayers with the same gross income: the first earns 90% of his revenues from matchable wage, whereas the second gets 90% of income from non-matchable sources. Clearly, it is materially possible for the latter to strategically conceal a vast amount of her income and devise profitable tax evasion. Fresh evidence from a study conducted in Denmark has indeed corroborated Francis Bacon's saying that *opportunity makes a thief*: tax evasion is found to be substantially higher for individuals who self-report their income compared to those who have most part of their resources reported by a third party to the tax agency, regardless of audits probabilities, [9].

Third, we allow population's risk aversion to be correlated with income and with the portion of matchable wealth. This allows for the customization of the audit policy to different states of the economy, as in booming phases the correlation of risk aversion and income in the whole population is likely to be negative, whereas the link may be weaker in gloomy periods.

Fourth, we test two ways to sense and adapt the perceived probability of being audited. Taxpayers can either estimate this probability using a sample average based on random matches with other peers or using their own history of past audits. We call these schemes *geographical* and *temporal* adaptation, respectively. It turns out that the efficacy of audit schemes critically depends on how agents subjectively estimate their audit probability in the next period.

On the one hand, taxpayers may, in this setup, evade less than theoretically predicted just because they cannot do otherwise. On the other hand, this model fully incorporates in the game a tax agency who is able to use some (minimal) pieces of information about the agents to maximize the revenues from its audit policy: this looks realistic, as more sophisticated schemes than purely random inspections are clearly within reach for a tax agency. Other works investigated some simple endogenous audit schemes [3, 6], finding significantly lower levels of underreporting, but letting the score depend on the degree of matchable income appears, to the best of our knowledge, novel and promising.

This paper is organized as follows. Section 2 provides details on the ways taxpayers and the tax agency are modelled. The following section discusses our simulation results and show how alternative audit policies interact with the distribution of matchable income, tax morale and agents' beliefs regarding the probability of verification. Section 4 provides some conclusive remarks and relates our findings with the existing literature, discussing a few policy suggestions.

2 The Model

We model a society made of a tax agency (TA) who collects taxes from N individuals. The next two subsections will provide a detailed description of the agents and of the society in which they are embedded. Such a society, as it is often the case in agent-based models, depends on some features of the tax system as well as on the meta-parameters of the distributions individual traits are drawn from.

2.1 Agents

Agents have exogenously given parameters, $I_j, \rho_j, \kappa_j, \beta_j, j = 1, \ldots, N$, held constant across time, which denote privately known income, risk aversion, tax morale and the fraction of matchable income, respectively. Individual parameters are sampled only once at the beginning of the simulation from a distribution described in the next subsection. Taxpayers decide what portion of their earnings d_{jt} ($0 \leq d_{jt} \leq 1$) to disclose at time $t = 1, \ldots, T$ in order to maximize their expected utility. The declared income, $d_{jt}I_j$ cannot fall short of their own matchable income $\beta_j I_j$ ($0 \leq \beta_j \leq 1$), which can proxy wage from (legal) employment, third-party reported income or earnings that cannot otherwise be concealed. Agents assume that they will be audited at time t with some probability p_{jt}, whose true value is unknown and must be estimated in one of two ways. The first method, called *geographical*, assumes that each taxpayer meets k "neighbors" per period[1] and learns the number m_t of those who were audited. The previously held $p_{j,t-1}$ is updated as follows:

$$p_{jt} = w p_{j,t-1} + (1-w) \frac{m_{jt} + A_{jt}}{k+1},$$

where $w = 0.5$ and A_{jt} is 1 if the j-th agent was audited at time t and 0 otherwise. In other words, the probability of experiencing an audit is the average of the past estimate, a fresh guess derived from the k encounters and the knowledge of whether she was audited herself. Alternatively, using *temporal* updating, each agent uses a time-average and computes, in any t

$$p_{jt} = \frac{1}{t} \sum_{i=1}^{t} A_{ji}.$$

While geographical updating almost exclusively depends on current audits on other taxpayers, temporal adjustment focuses only on the history of one agent, uses past information and is more accurate if a longer span of time is available.

The utility function of taxpayers is

$$U(d_{jt}) = (1 + d_{jt})^{\kappa_j} W_{jt}^{(1-\rho_j)}$$

where W_{jt} is the wealth after taxes and fines (if any) and κ_j ($0 \leq \kappa_j \leq 1$) represents the tax morale of agent j: the stronger the ethical sense and the more income the agent will report, the higher utility she will perceive. A stronger tax morale corresponds to a marked sensitivity of utility to changes in the choice of how much to conceal: indeed a less *moral* taxpayer, endowed with a lower κ_j, will feel less ashamed in not paying taxes than an upright one.

The fraction d_{jt} of income to be disclosed is optimally selected by each agent solving the problem

$$\max_{d_{jt} \geq \beta_j} EU(d_{jt}) = p_{j,t-1} U(X_{jt}) + (1 - p_{j,t-1}) U(Y_{jt}),$$

---

[1] We independently and uniformly sample the k neighbors from the whole population at each time and for each agent.

where $X_{jt} = (1 - \tau d_{jt})I_j - f\tau(1 - d_{jt})I_j$ is the income if audited and $Y_{jt} = (1 - \tau d_{jt})I_j$ is the income in the absence of an audit. In the previous equation, τ and f denotes the tax and penalty rates, respectively. Both parameters characterize the society in which agents live, which is described in the next subsection.

2.2 Society

We define "society" as the set of arrangements and parameters that describe the tax collection process and the distribution of individual traits previously detailed.

The TA collects taxes from agents setting the tax rate τ they must pay on their income and the fine rate f to be levied on hidden income, when and if evasion is discovered. The TA can audit a fixed number qN ($0 \leq q \leq 1$) of agents, according to some policy. We assume that this is done by assigning a score to each taxpayer and picking qN agents with probability proportional to the score. We consider four audit schemes: random auditing simply gives the same score to everyone in every period; strict cutoff sets the score to 1 for the qN agents whose declared income $d_{jt}I_j$ is the smallest and 0 otherwise; in the (mild) cutoff rule, auditing is performed proportionally to the rank of the declared income; finally, by enhanced auditing we refer to a scoring system that is developed by the TA in the attempt of maximizing the revenues from enforcement.

Using strict cutoff, the TA will audit in each period the qN individuals who declared the least. Agents who report low income are likely to be inspected also under cutoff auditing but there is much more variability with respect to strict cutoff and many more individuals experience one or more audits along time.

Enhanced auditing is based on the score S_{jt}, which is a function of matchable and declared income:

$$S_{jt} = p_1(\beta_j I_j)^{p_2} + (d_{jt}I_j)^{p_3},$$

where p_1, p_2 and p_3 are constants selected by the TA to approximately maximize the revenues from audits.[2] While random auditing is a special case of the enhanced scoring system (set $p_1 = 1, p_2 = 0, p_3 = 0$ to obtain constant scores), only a part of all possible functions of βI and dI is explored with this parametrization.

---

[2] The objective of the TA, the maximization of the sum of the fines imposed in qN audits, is a stochastic function of p_1, p_2, p_3 and the approximate solution depends also on the "givens" of the society. This is to say that different enhanced schemes are likely to be developed in different societies or when agents behave differently. We mimic the TA's search for good triplets of p_1, p_2, p_3 by means of an Evolution Strategies algorithm, which is stopped after 30 functions' evaluations and prematurely halts. Therefore, the process should be interpreted as a somewhat realistic quest by a boundedly rational TA of an audit scheme that is tailored to the society and capable of improving the revenues from fines. For additional details on Evolution Strategies see [4], for implementation details see [11].

Table 1 Description of the societal individual parameters' distributions

Name	Symbol	Distribution
Income	I	Lognormal, mean = 30,000, s. d. ≈ 23,500
Tax morale	κ	Uniform in $[\kappa_{low}, 1 - \kappa_{low}]$, $\kappa_{low} = 0, 0.025, \ldots, 0.25$
Risk aversion	ρ	Uniform in $[0, 1]$
Matchable income	β	beta(a, a), $a = 0.5, 1, 2$

Clearly, neither the strict nor the mild cutoff can be exactly replicated but the relative size of p_2 as compared to p_3 and the scaling factor p_1 can provide some guidance in singling which users are more likely to be audited using the enhanced scheme.

The description of the society is completed by the definition of the distributions used to draw the individual parameters, see Table 1. We assume population income to follow a log-normal distribution, with mean income being 30,000 and standard deviation equal to 23,500 (these figures vaguely reflect Italian ones); the risk aversion parameter is uniformly distributed in $[0, 1]$. For the sake of simplicity, we suppose that tax morale is uniformly distributed among the taxpayers and let the support of the density change symmetrically around the mean, thus effectively considering mean preserving "spreads" of the same distribution. This way we can easily represent different societies with distinct moral attitudes: on average tax morale has the same value in every country, but we can find societies with more extreme values – in both directions – than others and increments in κ_{low} increase aggregate compliance. The fraction of matchable income β_j for the j-th agent follows a beta(a, a) distribution and, in particular, we focus on three specific values of $a = 0.5, 1, 2$ describing density functions that are U-shaped, uniform and bell shaped, respectively. When $\beta \sim$ beta$(0.5, 0.5)$ most of the agents have either high or low matchable income, whereas there is low density for middle ways; on the contrary, a beta$(2, 2)$ distribution corresponds to a society where taxpayers mostly have a mix of matchable and non-matchable income, with few extreme cases. A country like Italy, where self-employment often leaves many opportunities for income disguising, can be representative of the first scenario ($a = 0.5$), whereas a nordic country, say Norway, where usually payments are completed by traceable means could better be approximated by the second situation ($a = 2$). The case relative to $a = 1$ stands in between.

Finally, we capture important second order effects in the distribution of citizens's individual traits of one society allowing for nontrivial correlation of parameters. Hence, while Table 1 reports the marginal distributions of parameters, we assume that $Cor(\rho_j, \beta_j) = r$ and $Cor(\rho_j, I_j) = -r$ across the population. Picking, say, $r > 0$ is tantamount to suppose that more risk-averse agents have on average a smaller income. At the same time, they are likely to have a larger matchable income. In most of the countries we can think of, this positive correlation appears to be reasonable to account for the self-selection process that generally leads risk-averse agents to seek employed job and risk-prone taxpayers to self-employ in more profitable activities that also leave more room for evasion.[3]

Since the TA does not make q public, taxpayers are not fully informed about the true intensity of audits and have a perceived probability of auditing p_{jt} that is updated according to one of the two possible ways explained before. Furthermore, taxpayers do not know how the tax agency actually runs the inspections and which algorithm is followed when selecting the tax files to audit.

3 Results

The model presented in the previous section generalizes the classic framework in several ways: agents are restricted in their compliance decision and must declare at least as much as their matchable income; they can update the perceived probability of audit using geographical or temporal adjustments, while the correlation among parameters and the level of tax morale of the population varies. We simulate a grid of beta distributions, where $a \in \{0.5, 0.75, \ldots, 2\}$ (7 values), and tax morale levels $\kappa_{low} \in \{0.025, 0.050, \ldots, 0.250\}$ (11 values). Each grid was then replicated four times, to account for two possible values if correlations among parameters, $r = 0$ and $r = 0.5$ (two values), and the two updating schemes, denominated geographical and temporal (two elements). For each set of values for the meta-parameters a, κ_{low}, r and one updating method, we simulated 30 periods with $N = 1,000$ agents, for a total of $7 \times 11 \times 2 \times 2 = 308$ societies, where $\tau = 27\%$ and $f = 1$ are constant. To avoid any dependence on the initial values of the probability p_{j0} at time 0, we discard the first 29 periods and report the results of the last one, which can be thought of as one fiscal year.

As customarily in agent-based models, the richness of the data is both a curse and a blessing and we especially focus in what follows on the audit policies, contrasting the random, enhanced, strict cutoff and mild cutoff audit systems. Figure 1 depicts the revenues from fines of the four policies, when agents geographically update their own subjective probabilities, $r = 0.5$ and $a = 0.5$.

The solid lines are relative to the gains of strict cutoff, enhanced and mild cutoff scoring, normalized by the revenues of the random scheme. For instance, with minimal tax morale ($\kappa_{low} = 0$), the strict cutoff and enhanced auditing produce revenues that exceed those of the random scheme by over 500% and about 350%, respectively, as can be seen in the left part of the picture. When tax morale is low, it is clear that audits based on the strict cutoff rule are more lucrative on average than

---

[3]Technically, we obtain correlated marginals as follows: we sample from a 3-dimensional multivariate normal with the given correlations; once we have a normal vector $(z_j^{(I)}, z_j^{(\rho)}, z_j^{(\beta)})$ for each agent, we invert the appropriate cumulative probability distribution (log-normal, uniform and beta, respectively) to obtain (I_j, ρ_j, β_j) with the desired approximate correlations.

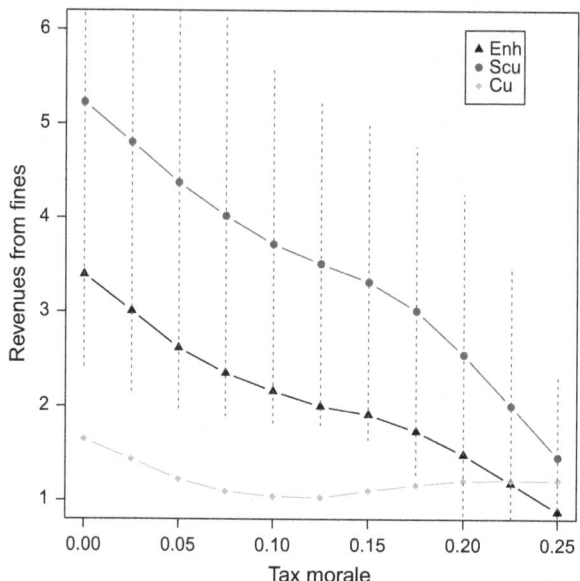

Fig. 1 Revenues obtained by strict cutoff, enhanced and mild cutoff selection methods (from top to bottom, normalized by the revenues of the random audit scheme), relative to geographical updating, $r = 0.5$, $a = 0.5$. The *vertical dashed segments* depict one standard deviation above and below the mean of revenues for the strict cutoff policy

any other scheme. Notice that this finding points out that the standard model, where only random auditing is performed and tax morale is identically null, is unstable in that the TA would have a strong incentive to change its auditing method. There is substantial variability in the outcomes, as displayed by the dashed lines showing one s.d. of revenues away from the mean: even though the average value is significantly higher under strict cutoff, there is a moderate and definitely non-null probability that, say, enhanced auditing will raise more fines in a single period (i.e., in a random batch of qN audits).

In the framework depicted in the figure, despite the effort made by the TA to "maximize" the revenues, the enhanced scheme is not the most profitable. This is due to the high level of noise present in the stochastic objective, to the limited resources allocated for the task (the search stops after 30 evaluations, see Footnote 2) and to specific features of the society.

The high average revenues generated by the strict cutoff rule derive from a somewhat extreme combination of effects: due to positive r, more wealthy taxpayers are likely to have lower risk aversion and are more inclined to evasion; in the cases where their tax morale and matchable income happen to be low, such agents simply conceal virtually all their income and declare $d_j \approx 0$. The strict cutoff rule often samples repeatedly wealthy total evaders and, therefore, revenues are boosted.

Tax Enforcement in an Agent-Based Model with Endogenous Audits

Table 2 Tax agency revenues from fines, when taxpayers *geographically* update p_{jt}. Values are normalized by the revenues of the random audit scheme

		$r = 0.5$		$r = 0$	
		LTM	HTM	LTM	HTM
$a = 0.5$	Enh	3.00	1.18	1.53	0.93
(U-shaped)	Scu	4.80*	2.00*	4.83*	1.75*
	Cu	1.43	1.21	1.19	1.00
$a = 1$	Enh	0.96	1.28	1.36	1.09
(Uniform)	Scu	3.53*	1.78*	3.15*	1.42*
	Cu	1.15	1.54	1.24	1.00
$a = 2$	Enh	0.85	2.61*	1.85	2.28*
(Bell-shaped)	Scu	1.76*	0.87	2.46*	1.25
	Cu	0.90	1.09	1.55	0.83

However, this is possible only when agents geographically update the probability of being audited and "do not realize" that they will probed much more frequently if they declare low incomes. For increasing levels of κ_{low} the difference in performance among the scoring rules markedly decreases, thus showing that revenues from tax enforcement are relatively insensitive to the audit policy in societies with high tax morale (besides obviously being much lower).

Table 2 shows a summary of our results when agents use geographical updating of the probabilities. The data in the table show how the relative revenues (with respect to random auditing) of the three considered policies vary as r, the tax morale, and the distribution of matchable income change in the society. We denote by LTM (HTM) low (high) tax morale societies in which $\kappa_{low} = 0.025$ ($\kappa_{low} = 0.225$).

The best performing audit policy (highlighted with *) is quite often the (rather brutal) strict cutoff selecting system. This always holds when tax morale is low, regardless of the distribution of the matchable income and for any value of r. The relative efficacy of the strict cutoff policy decreases moving from a U-shaped ($a = 0.5$) to a bell-shaped distribution ($a = 2$). This result is due to the reduction, as a moves from 0.5 to 2, of the number of "extreme" taxpayers who have, at the same time, low risk aversion, low individual tax morale κ_j and small matchable income. Hence, the strict cutoff rule increasingly audits real poors as opposed to aggressively non-compliant taxpayers.

The inspection of the columns labelled HTM, relative to high levels of tax morale, reveals that the enhanced policy obtains better results in the case of bell-shaped distribution of matchable income. The intuition here is straightforward: high tax morale and few individual with extremely low β requires a more nuanced audit policy, which does not concentrate itself on the smallest declaration but takes a more sophisticated approach in which both $d_j I_j$ and $\beta_j I_j$ have a role in the scoring function. This is confirmed by the values of p_1, p_2 and p_3 that determine the enhanced policy: for increasing a, p_2 and p_3 have the tendency to increase and

Table 3 Tax agency revenues from fines, when taxpayers *temporally* update p_{jt}. Values are normalized by the revenues of the random audit scheme

		$r = 0.5$		$r = 0$	
		LTM	HTM	LTM	HTM
$a = 0.5$	Enh	0.88	1.05	1.64	0.76
(U-shaped)	Scu	1.20*	0.31	2.02*	0.26
	Cu	0.83	1.24*	1.44	1.15*
$a = 1$	Enh	0.90	0.92	0.77	0.88
(Uniform)	Scu	1.13*	0.20	0.70	0.24
	Cu	1.09	1.31*	1.18*	1.29*
$a = 2$	Enh	1.53*	1.37*	0.69	0.78
(Bell-shaped)	Scu	0.84	0.27	0.55	0.25
	Cu	1.36	0.74	0.90	0.89

to cluster around zero, respectively, and this translates into a larger weight given to the βI term as compared to the dI one.[4]

Table 3 presents the simulation results when agents temporally update their subjective assessment of the auditing probabilities. Recall that in this case, taxpayers focus exclusively on their own audits' history, averaging over time the relative frequency of the inspections they have undergone. On the one hand, this disregards the information on the unconditional intensity of control q, which could be estimated by sampling information on the neighbors; on the other hand, however, any given taxpayer is likely to be able to estimate with much greater accuracy the probability of being audited conditional on his behaviour, which is ultimately what she should care about.

The insights that can be obtained by Table 3 remarkably differ from the ones descending from Table 2. We stress that this is only due to the different method in updating the subjective audit probabilities. Overall, as several values are smaller than 1, there are numerous cases in which random auditing over-performs at least one competing scoring scheme. When $r = 0$ and $a = 2$, surprisingly, random auditing appears to be the best policy. Generally, the difference in performance across policies is reduced under temporal updating and there is no clearly dominant strategy: strict cutoff is the best option in three cases (only), whereas enhanced scoring achieves the best results in two cases (for positive r and bell-shaped matchable income). Mild cutoff scoring secures the highest revenues in five cases, mostly related to societies with high tax morale. Evidently, as temporal updating allows taxpayer to partially anticipate the decisions of the TA, a good deal of randomness in the choice of the taxpayers to audit has favorable payoffs. Mild cutoff, indeed, can be interpreted as a random scheme endowed with some bias that increases audit probabilities for low declarations; indeed, as noticed before, this

---

[4] We thank an anonymous referee for this remark.

audit policy keeps sampling a variety of agents and does not get trapped in repeated audits of the same agents.

A careful scrutiny of the rich set of data available at the micro level shows an interesting dynamics going on between evasion-prone agents and the TA. A taxpayer subject to audit at time t will, under temporal updating, revise upward his belief about the probability $p_{j,t+1}$ of experiencing an audit in the following period. *Ceteris paribus*, this will increase the amount $d_{j,t+1}$ disclosed to the TA at $t+1$ and reduce the chance of an audit. As a consequence, $p_{j,t+2}$ would decrease on average, pushing the agents to conceal more income and – consequently – increase their likelihood of being audited at $t+2$ and so on. Such a *hide and seek* game is only possible when temporal updating is used by taxpayers and, in this setup, one of the best options for the TA would be, game theory *docet*, a good deal of randomization with an eye to low levels of *dI*, which is exactly what may be achieved by an enforcement policy guided by the mild cutoff scoring.

4 Conclusions

The agent-based model we have studied in this paper extends the standard model of tax evasion, allowing for heterogeneous taxpayers, consideration of matchable income, the adoption of several alternatives to the common random audit schemes, and two plausible ways to assess and update the subjective probability of being audited. Some (but not all) versions of the previous features were studied in previous works, but our model provides a comprehensive picture of the complex interactions occurring between the TA and the taxpayers. Such a detailed representation is precluded to most analytically solvable models, that have a more limited scope and must be based on simplifying (and often heroic) assumptions.

As a first general remark, the model shows that effective audit policies are dependent on the context. We confirm that the tax morale is an important factor in explaining tax evasion. Indeed, high values of tax morale, nearly always correspond to lower revenues for the TA (due to lower evasion rate). At the same time, the way agents perceive and update the probability of audit is also extremely relevant: if taxpayers take compliance decisions mainly based on audits experienced by others, there is scope for the TA to adopt the lucrative strict cutoff rule that simply audits those who declare the least. On the contrary, a seemingly minor modification in the way probabilities are perceived and updated, that is taking into account the history of each taxpayer, results in a different outcome where most enforcing policies are somewhat similar and mild cutoff, a biased version of random auditing, appears to be the best option to maximize revenues. Realistically, taxpayers will use a mix of the two stylized updating methods we considered but, even in this case, the TA should modify its actions depending on judgements or guesses about both the way agents behave and their tax morale. The temporal updating method is definitely more precise than the geographical one in determining the likelihood of being audited – when this is done not randomly – and, as a matter of fact, when agents become more acute, the TA needs to use more subtle selection methods. As people use geographical updating (and react too little to their own experience), the strict cutoff outperforms almost always the other methods. An exception is the case of beta(2,2) and high levels of tax morale, possibly because agents are willing to declare more (they are more moral) and there are less people with plenty of non-matchable income. The TA needs to be more sophisticated: metaphorically, when you go fishing in a sea with both big and small fish, a wide-mesh net can be used; but if you go to the fishing pond, where fish are not as big and maybe swim deep, then more ingenious means are needed.

Other structural features of the society, such as the distribution of the matchable income or the correlations among individual parameters, have relevant effects. A U-shaped configuration, where many agents have plenty of material opportunity for evasion, should be tackled by targeting low declarations (still keeping robust doses of randomness in the choice). In contrast, bell-shaped distribution of matchable income requires different audit policies and more nuanced approaches, giving more weight to the matchable component and considering the level of tax morale. There is even one case (see the bottom right panel of Table 3) in which the best policy is simply random auditing.

According to the simulation, the efforts exerted by the TA to develop enhanced auditing rule are, with the minimal information set at disposal here, of relative effectiveness in a dynamic setup. A bit paradoxically, audit schemes are working well if they are "deceptive": when taxpayers update their personal belief in a geographical way, implicitly assuming that the whole population is audited at random and each individual is equally likely to be picked, then the TA should optimally proceed with targeted audits, precisely because agents do not realize why they are chosen. Conversely, temporal update implies that taxpayers acknowledge that each individual is audited in way that reflects her individual features. In this situation, the introduction of some randomness on the part of the TA would shake the certainty of those who did not expected an audit and increase on average the efficacy of enforcement.

Acknowledgements We thank Matteo Richiardi and Dino Rizzi for useful discussions and suggestions. The financial support of PRIN 20103S5RN3 "Robust decision making in markets and organization" is acknowledged.

References

1. Allingham MG, Sandmo A (1972) Income tax evasion: a theoretical analysis. J Public Econ 1(3–4):323–338
2. Alm J, McClelland GH, Schulze WD (1992) Why do people pay taxes? J Public Econ 48(1):21–38
3. Alm J, Cronshaw MB, McKee M (1993) Tax compliance with endogenous audit selection rules. Kyklos 46(1):27–45

4. Beyer H, Schwefel H (2002) Evolution strategies – a comprehensive introduction. Nat Comput 1(1):3–52
5. Bloomquist KM (2003) Tax evasion, income inequality and opportunity costs of compliance. Paper presented at the 96th annual conference of the National Tax Association. Chicago
6. Collins JH, Plumlee RD (1991) The taxpayer's labor and reporting decision: the effect of audit schemes. Account Rev 66(3):559–576
7. Kahneman D, Tversky A (1979) Prospect theory: an analysis of decision under risk. Econometrica 47(2):263–292
8. Kirchler E, Hoelzl E, Wahl I (2008) Enforced versus voluntary tax compliance: the "slippery slope" framework. J Econ Psychol 29(2):210–225
9. Kleven HJ, Knudsen MB, Kreiner CT, Pedersen S, Saez E (2011) Unwilling or unable to cheat? Evidence from a tax audit experiment in Denmark. Econometrica 79(3):651–692
10. Torgler B (2002) Speaking to theorists and searching for facts: tax morale and tax compliance in experiments. J Econ Surv 16(5):657–683
11. Trautmann H, Mersmann O, Arnu D (2011) http://cran.r-project.org/web/packages/cmaes/cmaes.pdf

Subprime Lending and Financial Inequality in an Agent-Based Model

Andrea Teglio, Silvano Cincotti, Einar Jon Erlingsson, Marco Raberto, Hlynur Stefansson, and Jon Thor Sturluson

Abstract Real estate bubbles often trigger financial and economic crisis. U.S. subprime mortgage crisis and the Spanish property bubble, both occurring in 2008, are recent examples whose consequences are still affecting the respective economies. The aim of this paper is to understand if the level of concentration of financial capital has an impact on the real estate bubble formation. We study the issue in a first scenario where mortgage loans are easily granted (subprime mortgages) and in second one with a stricter regulation for the access to credit. Our results show that the combination of capital concentration and easy access to credit gives rise to a strong economic instability and to a highly unequal distribution of wealth.

1 Introduction

Recently, attention has been drawn to the problem of increasing inequality and the rising debt levels in developed economies. The Global Risks 2013 report of the World Economic Forum [6] rated income inequality as the most likely global risk to manifest sometime in the next 10 years. The effects of income inequality of households have been studied by e.g. [9], showing the negative effect it can have on economic growth.

A. Teglio (✉)
Departament d'Economia, Universitat Jaume I, Av. Sos Baynat, Castellón de la Plana, Spain
e-mail: teglio@eco.uji.es

S. Cincotti · M. Raberto
DIME, Università di Genova, Via Opera Pia 15, 16145 Genova, Italy

E.J. Erlingsson · H. Stefansson · J.T. Sturluson
Reykjavik University, Menntavegur 1, Reykjavik, Iceland

In many countries empirical models have been used for analysis and policy making and have become an established part of the policy framework [8]. The most used models are based on the general equilibrium framework (see e.g. [11]). Shortcomings of DSGE models and variations of such models have been known for a long time [7]. Recently, agent-based models have been proposed as an alternative tool for studying the effects of policy implementations within the economy [5]. Agent-based models have been promising when studying complex interactions (see e.g. [12]) and reproducing many stylized facts of the economy (see e.g. [3]).

The Iceace model, that is described in this paper, is an agent-based model of a credit-network economy. Various types of economic agents interact through various markets, e.g. consumption market, labor market, credit market and housing market using behavioral rules. The Iceace model is partly based on the EURACE model (see e.g. [10]), using, in particular, the balance sheet accounting of agents. A simulator has been developed from the model in order to study the workings of the artificial economy when it is subject to different policy settings.

2 The Model

The artificial economy is composed by households, firms, construction firms, banks, an equity fund, the government and a central bank. Households provide a homogeneous labor force to firms and constructions firms, buy homogeneous consumption goods produced by firms and new houses built by construction firms and exchange each other their stock of housing units. Banks supply loans to firms and construction firms, and provide real estate mortgages to households. The equity fund owns all the equity shares of firms, constructions firms and banks, collect their dividends and may provide liquidity to firms and constructions firms if needed. Shares of the equity fund are equally distributed among capitalists who are a fraction, κ, of households in the model. The government and the central bank perform respectively the fiscal and the monetary policy. The elementary time step of the simulation is conventionally set equal to a day; however, most of events occur on a weekly (consumption), monthly (production, production planning, labor and housing markets) on quarterly (financing, agents' income statement and balance sheet accounting) base.

2.1 Production and Pricing

Both firms and construction firms are characterized by a Leontief production technology with two inputs: labor units L and capital goods K, i.e.,

$$q^{(f,s)} = \min \left(\gamma_L^{(f,s)} L^{(f,s)}, \gamma_K^{(f,s)} K^{(f,s)} \right), \tag{1}$$

where $\gamma_L^{(f,s)}$ and $\gamma_K^{(f,s)}$ are the productivity of labor and capital, respectively; f or s are the indexes of the firm or the construction firm considered. The capital endowment should be considered as a constant as neither depreciation nor investments are considered.

In the case of firms, production takes place the last day of any month and is available for sale the first day of the following month. Conversely, in the case of construction firms, the production of each housing unit takes 12 months. Construction firms make a production plan $\tilde{q}^s$ at the beginning of any month regarding the number of housing unit projects to advance during the month and decide to increase (decrease) by a random amount the housing unit projects to advance depending on the increase (decrease) in hosing price during last month. Therefore, ongoing housing unit projects can be stopped[1] because of the new plan or also because the construction firm is rationed in the labor market or it is forced to layoff employees as it is unable to collect funding in the credit market or from the equity fund. Only when a construction project has been advanced 12 times (i.e. for not necessarily consecutive 12 months), then the housing unit is complete and is available for sale.

Firms perform a fixed mark-up pricing on unit costs given by unitary labor costs and unitary financing costs. Construction firms set sale prices in the housing market according to the last month market price, see Sect. 2.3 for details.

Firms make production plans $\tilde{q}^f$ at the beginning of each month for the present month. Firms production plans depend on the sales expectations $\hat{q}^f$ in the next month and on the inventory level I^f. We assume that sales expectations are equal to previous month sales q^f except if in the previous month all the inventories were sold out. In this latter case, expected sales are set equal to an amount 10% higher than sales in the previous month. In the general unsold inventories case, the production plan $\tilde{q}^s$ is then set to:

$$\tilde{q}^f = \eta q^f + (1 - \eta)\left(\hat{q}^f - \max(I^f - \hat{q}^f, 0)\right). \qquad (2)$$

Equation 2 takes into account a weighted average, with weight $\eta \in (0, 1)$, between previous month production q^f and the supposed optimal plan, i.e. $\hat{q}^f - \max(I^f - \hat{q}^f, 0)$. The weighted average takes into account a realistic inertia on previous productions and allows to avoid too wide output oscillations. The rationale of the supposed optimal plan is to produce the expected sale foreseen next month $\hat{q}^f$, as production will be only available after 1 month, minus the possible remaining inventories unsold in the present month, considering that $\hat{q}^f$ will be also the amount sold this month. Given $\tilde{q}^s$ and $\tilde{q}^f$, construction firms and firms compute the labor force L_n^f needed to fulfill their plans as:

---

[1]The construction project stopped are the less advanced ones.

$$L_n^{(f,s)} = \frac{\tilde{q}^{(f,s)}}{\gamma_L}.\tag{3}$$

The labor market is decentralized and is active the first day of any month after firms and construction firms set their production plans and then their needed labor force according to Eq. 3. The difference between the needed labor force and their present labor endowment determines the labor demand and then new job vacations (if positive) or layoffs (if negative).

The labor market is characterized by four phases. First, producers with a positive labor demand increase their wage offer by a fixed percentage to keep their present workers as well as to attract new ones and post open job positions. Second, producers with a negative labor demand fire workers[2] in excess. Third, some employees are randomly selected and queued according to their skills to look for new job positions which they take if the offered better wages. Finally, unemployed households are queued according to their skills to fill the remaining, if any, open positions.

2.2 Financing and Bankruptcy

Banks provide loans to firms and construction firms to finance their operations and mortgages to households willing to buy housing. Banks are constrained in their lending activity by a Basel II-like minimum capital requirement rule concerning a minimum fraction between their equity and the sum of their risky assets, i.e. loans to firms and mortgages to households.

Loans are infinitely lived and borrowers need to pay back on a quarterly base only interests. Conversely, mortgages have a life span of 40 years and households have to pay back both interests and fractions of the principal. In particular, we consider adjustable-rate mortgages (ARMs) where the mortgage rate r_M is given by the central bank interest rate plus a fixed 2 % spread. The repayment costs R^h of any mortgage are paid by household h on a quarterly base and include both interests, R_r^h, and the repayment of a part R_U^h of the to-date mortgage principal U^h. In the ARMs case, total mortgage costs are computed as $R^h = U^h/A$ where A is the annuity factor,[3] the quarterly interest payment is simply given by $R_r^h = U^h r_M/4$ and the part repaid of the principal is given accordingly by a residual, i.e., $R_U^h = R^h - R_r^h$.

Firms must have also a positive equity to receive a loan, while households to get a mortgage must have a minimum equity ratio and need to show to be able to pay the costs (interests + principal repayment) of all their mortgages, including the

---

[2]The households with the lowest skills are selected for firing.
[3]The annuity factor is generally computed considering the remaining life of the mortgage as $A = \frac{1}{\frac{1}{4}r_M} - \frac{1}{\frac{1}{4}r_M\left(1+\frac{1}{4}r_M\right)^n}$ where n is the number of remaining quarters in the mortgage life.

new one, given their present income and the present mortgage rate. In particular, the sum of quarterly costs of any present mortgage i, i.e. $\sum_i R_i^h$, plus the additional quarterly costs of the new requested mortgage i^*, i.e. R_{i*}^h, must not be higher than a fraction β of the total quarterly net income, including both labor Z_ℓ^h and capital income Z_e^h. The condition that needs to fulfilled by a household to get a mortgage is then:

$$\sum_i R_i^h + R_{i*}^h \leq \beta \left(Z_\ell^h + Z_e^h \right). \tag{4}$$

The market for loans to firms and construction firms opens each quarter and is active the first day of the period. Both types of producers do not make capital investment, however demand for loans may arise to meet liquidity needs to pay dividends to the equity Fund and interests to banks. Producers and households have a preferred bank whom to apply for borrowing, but if rationed due to insufficient capital requirement of the bank, only firms and constructions firms have the chance to apply for a second time to a different randomly selected bank. If still rationed, both types of producers have the opportunity to reduce dividends payment to zero, but if this is still insufficient to meet the liquidity needs to pay loan interests, they can request financing in the form of equity to the Equity Fund if their equity ratio is above the minimum required by the Equity Fund which is 5%. If this condition is not fulfilled, then the producer goes bankrupt (illiquidity bankruptcy) and the producer's debt is restructured so that the payment of interest is equal to the last quarterly earnings before interest). In the case of negative earnings before interests the debt of the firm will be set to zero. A second more severe form of bankruptcy (insolvency bankruptcy) occurs when a firm or a construction firm results to be characterized by negative equity at the quarterly balance sheet accounting. In this case, all the producer debt is written off, causing a severe loss for the banking system, and all its employees are laid off. A new producer is then started with one employee and initial physical capital endowment inherited from the failed producer.

2.3 Housing Market

The housing market is active the first day of any month and is a decentralized posted-price market where sellers post prices and buyers search for the cheapest housing units; households can buy or sell one housing unit at any market round (month) and housing units are homogeneous. With respect to an earlier version of the model in [4], constructions firms participate in market by selling new produced housing units. Beside construction firms, market activity is characterized by households randomly selected to enter the market as buyers or sellers with equal likelihood and by fire sales by households subject to financial distress that are forced to sell housing to reduce their mortgage burden. The condition for triggering a fire sale is when households are subject to quarterly mortgage costs (interest + principal payments) R^h higher than a given fraction θ of their total quarterly net income, given by both labor income Z_ℓ^h and capital income Z_e^h. Sellers post prices which are based on the last market round (last month) average transaction price P_H. In particular, households selected to be random sellers and construction firms post selling prices higher than P_H by a random amount as we assume that, in the absence of financial distress, sellers are willing to sell their housing units with the condition to make a gain with respect to the reference price P_H. Conversely, in the fire sale case, sellers are supposed to have the necessity to sell and therefore post selling prices lower than P_H by a random amount. Buyers are randomly queued and at their turn buy the cheapest available housing unit provided that they have the necessary financial resources or get a mortgage from a bank. The market closes when all buyers have their turn or the supply of housing is fully depleted. A new housing price P_H is then calculated as the average of realized transaction prices.

2.4 Consumption

We assume that any household h sets its monthly budget for consumption C_B^h depending on its monthly disposable income Υ^h and on the wealth effect caused by an increase or decrease in its net wealth, E^h, mainly due to rising or falling housing prices.[4] The monthly disposable income is given by $\Upsilon^h = \frac{1}{3}(Z_\ell^h + Z_e^h - \sum_i R^h)$, i.e. by the sum of the labor Z_ℓ^h and capital Z_e^h income realized in the last quarter minus the cost of all mortgages, i.e. $\sum_i R_i^h$. Consumption decision then is mainly modeled according the theory of buffer-stock saving behavior [2], which states that households consumption depends on a precautionary saving motive, determined by a target ratio ρ_C between the stock of liquid wealth, M^h, and the monthly disposable income, Υ^h. The monthly consumption budget C_B^h is then given by:

$$C_B^h = \Upsilon^h + \alpha_C (M^h - \rho_C \Upsilon^h) + \omega \Delta E^h. \qquad (5)$$

The rationale of this rule is to adjust the consumption budget every month so to adaptively meet the pre-determined liquid wealth to disposable income target ratio by consuming less (more) than the disposable income if $M^h < \rho_C \Upsilon^h$ ($M^h > \rho_C \Upsilon^h$) so to increase (decrease) M^h. The parameter α_C sets the speed of adjustment. The additional term $\omega \Delta E^h$ captures the wealth effect of housing and the parameter ω sets its size on consumption with respect to the net wealth quarterly variation ΔE^h.

The consumption market opens at the beginning day of every week and all households are randomly queued to select a firm whom to buy consumption goods with a probability inversely proportional to the price. The consumption goods market closes either when there are no goods for sale or when all households have spent their entire weekly consumption budget.

---

[4] According to [1], this wealth effect ranges from 5 to 8%.

2.5 Policy Making

Monetary and fiscal policies are performed respectively by the central bank and the government. The Central Bank sets its policy rate on a quarterly base according to a Taylor rule which takes into account both the rate of inflation and the unemployment rate, as a proxy of the output gap. The central bank acts as a liquidity provider for the banking sector.

The government collects taxes on households' labor and capital income and provides unemployment subsidies to unemployed households and general welfare benefits to all households. The government will always aim to have zero deficit, thus raising (lowering) taxes and lowering (raising) benefits if the balance is negative (positive).

3 Results

In this section we present a computational experiment designed to analyze the impact of a different percentage of capitalists on the economic outcomes of the model. Each capitalists' percentage is tested in two different contexts. A first case where banks grant mortgages to households only if they fulfill the requirement of being in a robust financial condition, and a second case where these requirements are relaxed (subprime lending).

In particular, the parameter β sets banks' criteria w.r.t. evaluating the eligibility of households for a mortgage (see Eq. 4). A higher β means looser creditworthiness conditions required by the bank to grant a mortgage. Parameter κ defines the percentage of total households which hold the equity fund's shares and which are therefore entitled to receive dividends paid by firms and banks. We test the impact of four potential scenarios, depending on the parameters κ and β.

The simulations are run with the following number of agents: 8,000 households, 125 firms, 25 construction firms, 2 banks, 1 government and 1 central bank. For each configuration of the parameters, several random seeds, generating different stochastic processes, have been used in order to improve the reliability of the outcomes. All the simulations are initialized with the same set of parameters, with the exception of β and κ. Several time series of a representative simulation are shown in Figs. 1 to 4. Apart from Fig. 1, where the four cases are plotted together, all figures are divided into two subplots, the upper one representing the case of stricter households' requirements, i.e., $\beta = 0.25$, and the lower one representing the case of looser requirements, i.e., $\beta = 0.3$. Each of the subplots shows two time series corresponding to a capitalists' percentage of 30 % (the dotted line) and 10 % (the thin line).

Figure 1 shows a clear ordering in the distribution of households' equity. As could be expected, a more unequal distribution of the equity funds' shares leads to a more unequal distribution of households' equity. Moreover, the easy access to

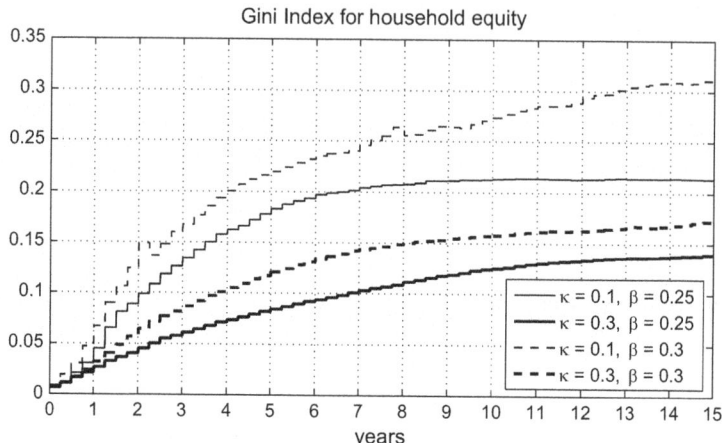

Fig. 1 Gini index computed on households' equity

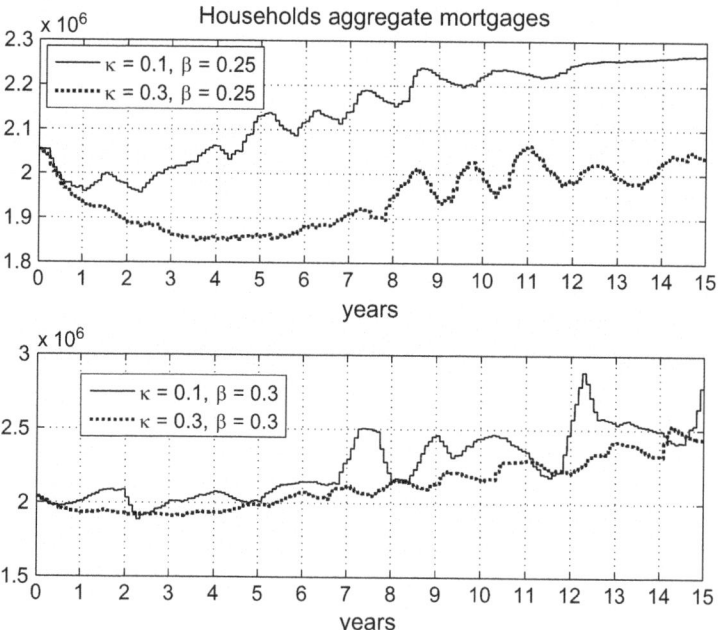

Fig. 2 Aggregate mortgages in the four scenarios

credit (subprime lending case, $\beta = 0.3$) also determines an unequal distribution of households' equity with respect to a stricter mortgage regulation.

Comparing the two cases of a high and a low percentage of capitalists among households, some basic differences can be pointed out. The housing market is more

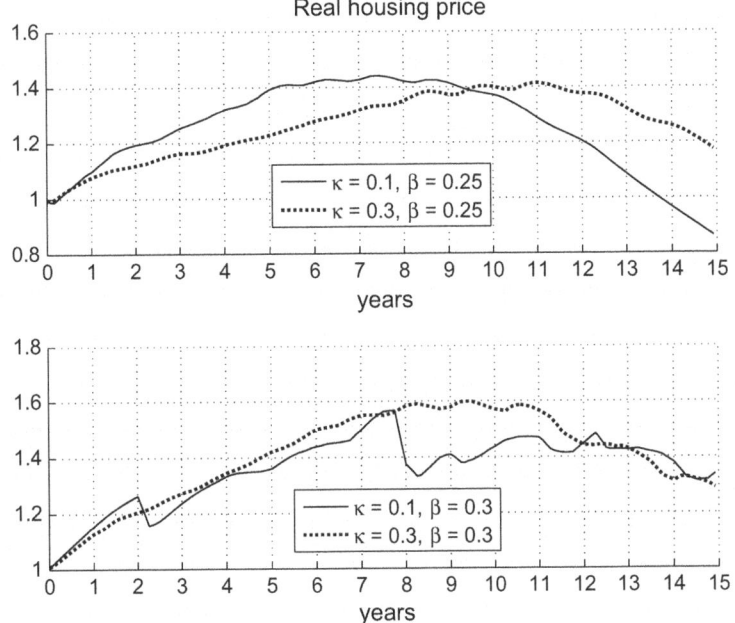

Fig. 3 Housing price level in the four scenarios

active when few capitalists own all companies' shares. They have high financial incomes and they are therefore able to get mortgages. Figures 2 and 3 show that both the amount of granted mortgages and the housing price grow faster in the case of a more concentrated financial capital.

However, while the mortgage and price growth seems to be sustainable in the case of a strict bank's lending regulation (low β), it causes an early crisis in the case of subprime lending (high β). Many households are not able to pay back their mortgages and are forced to sell their housing units, through a fire sale of housing, at a lower price, triggering the bubble crash and a consequent loans write off which can be observed in Fig. 2. Let us point out another detail: when the write off occurs, at year 2, there is a significant increase in the Gini index. This can be observed also in later occurrences, meaning that loans' write off generally imply a sudden growth in wealth inequality. The reason is that loans are mostly granted to capitalists and when the bubble crushes, and the housing price falls, they benefit from the debt write off more than non capitalist households, who are less indebted. When, at year 2, the housing price bubble bursts, as shown in Fig. 3, a debt deleveraging process (Fig. 2) severely affects the GDP (Fig. 4).

A clear feature that emerges from the scenario of subprime lending with a low capitalists percentage is the high volatility of the economic indicators. This can be seen in the time series presented in the figures and also in Table 1. The housing market is characterized by phases of intense activity, represented by the

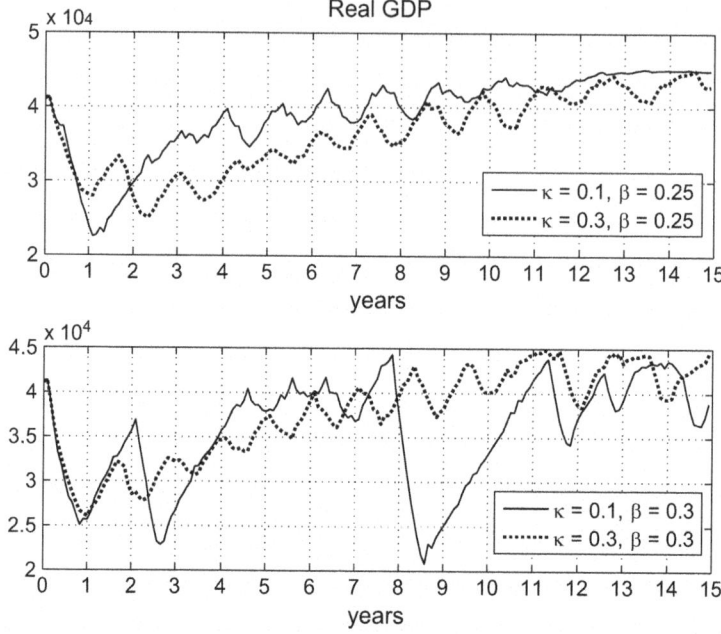

Fig. 4 GDP time path in the four scenarios

Table 1 Average values of some economic indicators of the simulated economy. Standard errors are in parenthesis

	$\beta = 0.25$ $\kappa = 0.1$	$\beta = 0.25$ $\kappa = 0.3$	$\beta = 0.3$ $\kappa = 0.1$	$\beta = 0.3$ $\kappa = 0.3$
Real GDP	38900 (7636)	36146 (6245)	35120 (5255)	36481 (5738)
Yearly GDP growth (%)	2.28 (3.64)	1.86 (1.78)	2.05 (2.16)	1.43 (1.68)
Variance of yearly GDP growth (%)	0.14 (0.08)	0.05 (0.01)	1.89 (0.62)	0.66 (0.32)
Housing price	192.2 (14.6)	188.5 (17.7)	206.2 (17.7)	212.5 (20.2)
Household mortgages	2163961 (38477)	1946054 (24288)	2202218 (50481)	2153601 (65254)
Firm leverage	2.29 (0.12)	2.48 (0.12)	3.59 (0.25)	2.91 (0.11)
No. firm bankruptcies	0.0 (0.0)	0.0 (0.0)	5.1 (2.3)	0.7 (0.5)
No. fire sales	1.2 (2.1)	0.5 (1.3)	50.1 (31.2)	7.9 (7.5)

raise of mortgage loans, and phases of debt deleveraging. A GDP depression reduces the economic activity in the housing market, due to the high unemployment which affects labor income and to the reduced level of production and profits, affecting financial income. However, after the recovery process, partially financed by a persistent deficit spending run by the government, the easy access to credit makes the housing market ready for a new bubble. Somewhat before year 8, a new and deeper crisis knocks down the economic system. Again, the trigger is the bubble in the housing market, generated by the excessive debt level. This time the depression is longer, during around 3 years. Again, the recovery process is driven by a fiscal expansion run by the government in order to support the aggregated demand, mainly paying unemployment benefits. It is worth noting that a new mortgage bubble starts just 1 year after the crisis, at the beginning of the recovery process. This is due to the relaxed regulation of the subprime lending scenario, where households can basically get a mortgage also when they are unemployed, with obvious implications on the economic stability.

An important role for determining the development of the economic crisis is played by the requirement on capital adequacy ratio (CAR), i.e., ratio of a bank's capital to its risk weighted assets. According to the Basel II regulation, a commercial bank should have a minimum CAR in order to guarantee adequate robustness and solvency. If the equity capital of the bank falls below the required level, the bank can not grant new loans or add any other risky asset to its balance sheet; see [10] and [12] for a study on CAR in an agent-based context. In the current setting, the minimum CAR is fixed at 5 %, a quite realistic value. In the simulations of Figs. 1–4, the CAR does not fall below the critical level and banks are never prevented from granting loans according to the Basel II requirements. If they do not grant a loan, it is only because of an unsatisfactory creditworthiness of borrowers. However, this is not always the case. In Fig. 5 we show another simulation where the Basel II regulation is infringed and banks are not in the condition to grant new loans. The effect is a long debt deleveraging period with a very slow recovery of GDP, mainly given by the increased public spending.

In Table 1 the average values of several economic indicators are presented for each of the four different scenarios; ten random seeds have been used. Results are in line with the analysis based on the presented figures. The case of low requirements and concentrated capital is clearly the most unstable economic scenario. Unemployment rate is high, GDP level is low, GDP growth is weak and GDP instability is high. Firm and households total debt is also bigger that in any other case, and the housing market is very unstable and characterized by a huge amount of fire sales.

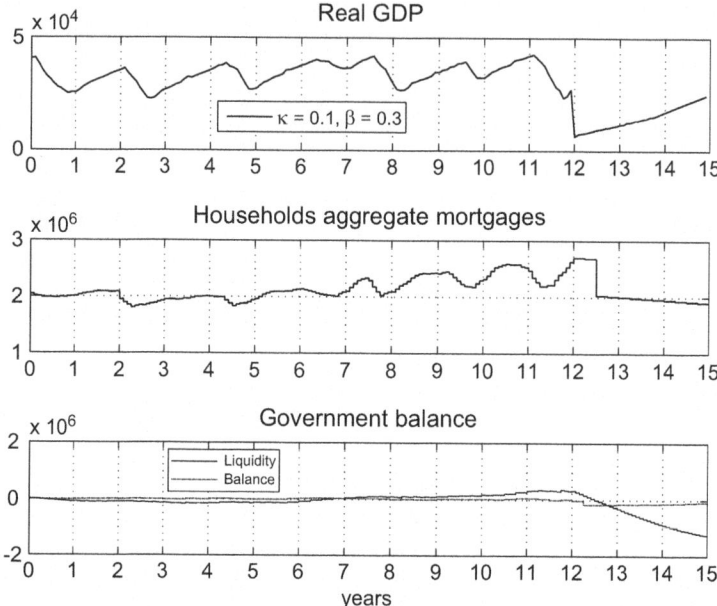

Fig. 5 The effect of Basel II constraint on capital adequacy ratio

4 Concluding Remarks

In general, when households requirements for getting a mortgage are stricter the economic performance is better and more stable. However, if requirements are too strict the economic performance is poor; this observation is not corroborated by the data presented in this paper, but emerges from a larger set of simulation performed.

Concerning the different impact of a more diffused or concentrated financial capital, results show that in the case of subprime lending a concentrated capital has very negative effects on economic stability and growth. When households' requirements are strict enough, the concentration of financial capital seems to improve economic indicators, although wealth distribution becomes much more unequal.

Acknowledgements The authors would like to acknowledge the financial support of the Icelandic Center for Research (RANNIS), grant no. 110653021. AT also acknowledges projets GV/2012/045 of the Generalitat Valenciana, project Bancaja P11A2010-17, and the Spanish national project ECO2011-23634.

References

1. Calomiris C, Longhofer S, Miles W (2012) The housing wealth effect: the crucial roles of demographics, wealth distribution and wealth shares. Technical report, National Bureau of Economic Research, Cambridge, MA
2. Carroll CD (2001) A theory of the consumption function, with and without liquidity constraints. J Econ Perspect 15(3):23–45
3. Dosi G, Fagiolo G, Napoletano M, Roventini A (2013) Income distribution, credit and fiscal policies in an agent-based Keynesian model. J Econ Dyn Control 37(8):1598–1625
4. Erlingsson E, Raberto M, Stefnsson H, Sturluson J (2013) Integrating the housing market into an agent-based economic model. In: Teglio A, Alfarano S, Camacho-Cuena E, Ginés-Vilar M (eds) Managing market complexity. Lecture notes in economics and mathematical systems, vol 662. Springer, Berlin/Heidelberg, pp 65–76
5. Farmer J, Foley D (2009) The economy needs agent-based modelling. Nature 460(7256):685–686
6. Howell L (2013) Global risks 2013, 8th edn. Technical report, World Economic Forum, Cologny/Geneva
7. Kirman A (1989) The intrinsic limits of modern economic theory: the emperor has no clothes. Econ J 99(395):126–139
8. Meen D, Meen G (2003) Social behaviour as a basis for modelling the urban housing market: a review. Urban Stud 40(5–6):917–935
9. Persson T, Tabellini G (1991) Is inequality harmful for growth? theory and evidence. Technical report, National Bureau of Economic Research, Cambridge, MA
10. Raberto M, Teglio A, Cincotti S (2012) Debt, deleveraging and business cycles: an agent-based perspective. Econ – Open-Access, Open-Assess E-J 6(27):1–49
11. Smets F, Wouters R, Europeo BC (2002) An estimated stochastic dynamic general equilibrium model of the Euro area. European Central Bank, Frankfurt
12. Teglio A, Raberto M, Cincotti S (2012) The impact of banks' capital adequacy regulation on the economic system: an agent-based approach. Adv Complex Syst 15(supp02):1250040

Part III
Market Dynamics

Adaptive Trading for Anti-correlated Pairs of Stocks

Chih-Hao Lin, Sai-Ping Li, and K.Y. Szeto

Abstract The effect of anti-correlation between stocks in real stock market can be exploited for profit if one can also properly set the criterion for trading that takes into account the volatility of the stock pair. This complex problem of resource allocation for portfolio management of stocks is here simplified to a problem of adaptive trading with an investment criterion that evolves along with the time series of the stock data. The trend of the stock is modeled with standard stochastic dynamics, from which the volatility of the stock provides a criterion for investment on a two stock portfolio that consists of the anti-correlated pair using mean variance analysis that optimizes the return. The action of buy and sell of the two-stock portfolio will be based on the fractional return of the pair: when the fractional return of the pair is greater than an upper threshold of 1.01, the action "buy" is taken; and when this fractional return is less than a lower threshold of 0.99, the action "sell" is taken. Since both the volatility criterion for investment and the fractional return of the two-stock portfolio are time dependent, the entire trading scheme is adaptive. Comparison of this evolving strategy of investment with time-average performance of the respective stocks indicates a consistent superiority.

C.-H. Lin (✉) · S.-P. Li
Institute of Physics, Academia Sinica, Nankang, Taipei 115, Taiwan, China
e-mail: chlin1019@phys.sinica.edu.tw; spli@phys.sinica.edu.tw

K.Y. Szeto
Department of Physics, The Hong Kong University of Science and Technology, Clear Water Bay, Kowloon, Hong Kong SAR, China
e-mail: phszeto@ust.hk

1 Introduction

Financial time series are intrinsically non-stationary, making their prediction and subsequent resource allocation for investment an extremely challenging task. The traditional theory of mean variance analysis by Markowitz [1] provides an intelligent guidance in portfolio management, but time dependent resource allocation in the context of financial portfolio management remains a difficult problem of continuous interest. Markowitz's analysis produces a simple and elegant solution for resource allocation, such as in the fraction of money invested in each constituent stock in the two-stock portfolios by specifying the investment frontier and the risk tolerable by the investor, which is different for different people. In reality, the time series is non-stationary and the static picture of the mean, the variance and the cross correlation are always changing. Thus, time dependent resource allocation remains an active field [2–7]. A standard approach is time series analysis, for example, through pattern recognition [8,9], genetic algorithm [10–12], neural network [13], or even fuzzy rule [14, 15].

One recent approach for extending Markowitz analysis is to make the portfolio composition a time dependent one, making use of both the long time behavior of the stocks as well as their short time fluctuation [16, 17]. This approach shows that there exists portfolio with low risk and high return, in spite of the random nature of the stock price and the unknown mechanism between the price variations of individual stock. However, this two time scales analysis involves large computational effort as it also provides possible switching between various two-stock portfolios. It will be nice to obtain some simple adaptive trading scheme that captures both the slowly changing volatility and correlation and the medium term fluctuation in the price of the two-stock portfolio. By the term "adaptive trading scheme" we imply setting up a set of trading criteria that evolves with the financial time series. Although this seems difficult, we will show in this paper a simple two-step filtering process for investment, with a filtering system that is co-evolving with the input data. In order to define our algorithm in the context of the idea of long and short term, we must introduce various parameters to define the time window for long, medium and short term. Based on a reasonable choice of these parameters, we succeed in comparing the performance of our adaptive trading scheme with the collective return of the stocks. The results are encouraging as our numerical analysis indicates consistent superiority of the adaptive trading scheme, in bullish or bearish market alike.

We handle the non-stationary nature of the financial time series in a different way. We first use the traditional tool of stochastic dynamics to create a criterion for investment window. Only when the online data satisfy certain condition based on the volatility ratio can we consider possible buy and sell action of the two-stock portfolio. Next, we use a simple version of mean variance analysis to compute the fractional change of the value of the portfolio over a relatively medium term period. If the fractional value satisfies a second criterion, we will then take action, either buy or sell, on this portfolio. The computation process is simpler here as we do not consider all possible pairs of stocks in the market, but only among three pairs that are preselected based on their anti-correlation. We select those pairs that are most negatively correlated, as they will provide the highest return per unit risk according to mean variance analysis. We will now describe our two methods of setting up the two criteria for investment with adaptive trading.

2 First Criterion on Investment

In stock markets, the stocks and their indices are interdependent and most of them are positively correlated. If the index increases, the stocks have higher probability to rise, which means if we can predict the trend of the index, we can also predict the stocks' behavior. By considering both the mean variance analysis and the trend prediction, we propose a transaction algorithm to reduce the transaction risks and increase the profits. We first select a pair of anti-correlated stocks to form a two-stock portfolio. The main reason to consider pairs of anti-correlated stocks instead of individual stocks in our trading algorithm is to reduce risk, as indicated by the existence of investment frontier in the Markowitz theory of portfolio management [1]. We then use the standard tool of stochastic dynamics to describe the trend of the stock market.

In mathematical finance, researchers usually assume that stock markets follow some stochastic processes, with stochastic differential form:

$$dS_t = rS_t dt + \sigma S_t dW_t, \qquad (1)$$

where r is the instantaneous risk free rate, σ_t is the instant volatility, and W_t is a Wiener process. We define the instant volatility as:

$$\sigma_t = \frac{1}{N_S - 1} \sum_{i=t-N_S+1}^{t} \left(\frac{dS_i}{S_i} - \left\langle \frac{dS_i}{S_i} \right\rangle \right)^2, \qquad (2)$$

where the N_S is sample size, which is chosen to be 30 days here. The trend of stock market is highly related to the volatility of the index: the crash of the stock market usually occurs more rapidly than its rise and the volatility in bearish market is in general higher. Real data also show that when the instant volatility rises, the index for the stock market has a higher probability to fall. Thus, we may use these observations to formulate a criterion for investment. First, we should consider the average instant volatility over a longer period. Let us assume that the average is taken from the previous year to the present and we denote the average instant volatility by $\langle \sigma_t \rangle$. Now that we have an average over a long term, (1 year), we can consider the fractional change of the instant volatility. The average $\langle \sigma_t \rangle$ can now act as an indicator to detect if the instant volatility is too high or too low. There are now four cases in comparing σ_t with $\langle \sigma_t \rangle$.

1. $\sigma_t > \langle\sigma_t\rangle$ and the stock market is rising,
2. $\sigma_t > \langle\sigma_t\rangle$ and the stock market is falling,
3. $\sigma_t < \langle\sigma_t\rangle$ and the stock market is rising,
4. $\sigma_t < \langle\sigma_t\rangle$ and the stock market is falling.

First let us be a conservative trader and assume that if $\sigma_t > \langle\sigma_t\rangle$, the stock market is likely going to fall, which agrees with empirical observation. Thus, when $\sigma_t > \langle\sigma_t\rangle$, the transaction is forbidden. However, if our assumption happened to be wrong, and the stock market's behavior is indeed case (1), we will then miss a chance of making profit. To remedy this, we calculate the average of $\frac{dS_t}{S_t}$ of the index over a medium period, for example, in the past 30 days, and denote it by $\left\langle\frac{dS_t}{S_t}\right\rangle$. If it is positive, the future trend of the index will then have higher probability to rise. Therefore, we modify our criterion for investment as follows: if $\sigma_t > \langle\sigma_t\rangle$, the transaction is forbidden except when $\left\langle\frac{dS_t}{S_t}\right\rangle$ is bigger than 0.

3 Stock-Pair Selection: Variance Analysis for the Long Time Scale

In mean-variance analysis for two-stock portfolios, we need first to define the mean $U(t)$ and variance $Var(t)$ as

$$U(t) = \frac{1}{N_S - 1} \sum_{k=t-N_S+1}^{t} p(k), \quad (3)$$

$$Var(t) = \frac{1}{N_S - 1} \sum_{k=t-N_S+1}^{t} (p(k) - U(t))^2, \quad (4)$$

where $p(t)$ is the daily closing price of the stock. The sample size, N_S, is again chosen to be 30 days. In our study of a two-stock portfolio, the variance for stock pair, s_i and s_j, is

$$var(s_i, s_j) = var(s_i)x_i^2 + var(s_j)x_j^2 + 2cov(s_i, s_j)x_i \cdot x_j, \quad (5)$$

where x_i and x_j are the fraction of the portfolio invested in $stock_i$ and $stock_j$ respectively. Note that the constraint $x_i + x_j = 1$, with $x_i, x_j \in (0, 1)$. The covariance $cov(s_i, s_j)$ is

$$cov(s_i, s_j) = \frac{1}{N_S - 1} \sum_{k=t-N_S+1}^{t} (p_1(k) - U_1(t))(p_2(k) - U_2(t)) \quad (6)$$

To avoid situations when the variance of one stock dominates the combined variance by improper selection of x_i and x_j, we choose

$$x_i = \frac{\sigma_{s_j}}{\sigma_{s_i} + \sigma_{s_j}}, \qquad (7)$$

$$x_j = \frac{\sigma_{s_i}}{\sigma_{s_i} + \sigma_{s_j}}, \qquad (8)$$

where σ_{s_i} is the standard deviation of stock s_i.

Prior to transaction, we select 8 stocks and sort them according to their variances in descending order which are named from stock1 to stock8. We then select three stock pairs: one from stock1 to stock4, one from stock3 to stock6, and one from stock5 to stock8. Each stock pair consists of $stock_i$ and $stock_j$, and it has the most negative correlation coefficient in its own group. This can reduce the risk during the investment.

We divide our fortune into three equal portions, and put them into each group, which follows the investment weight of x_i and x_j, corresponding to the stocks s_i and s_j. Therefore, we can reduce our risk by distributing the fortune into three portions of most negative correlated stock-pairs.

4 Short Term Trading Criterion

Apart from the long term trading criterion, it is necessary to also include the short term criterion. In previous section, we have one stock pair from each group, and the present price of this pair is denoted by $P_t = x_i \cdot stock_i(t) + x_j \cdot stock_j(t)$. We compare this portfolio with its average price from previous trading day t_0 to present time t: $\langle P_t \rangle = x_i \cdot \langle stock_i(t) \rangle + x_j \cdot \langle stock_j(t) \rangle$. If the ratio of $\frac{P_t}{\langle P_t \rangle}$ is larger than the threshold, we are going to buy this portfolio and vise versa.

There are two kinds of threshold, namely, fixed threshold and adaptive threshold. The fixed threshold means that the threshold is a constant during the whole trading period, which is set to be 1.01 here. However, the fixed trading threshold is not suitable for the whole period. The threshold should be lower for the rising trend of the index and higher for the falling trend of the index. We propose here the adaptive threshold and for our study here, the rule is as follows:

1. If $\langle \sigma_{t-1} \rangle > \sigma_{t-1}$, the adaptive threshold is 1.01,
2. If $\langle \sigma_{t-1} \rangle < \sigma_{t-1}$, the adaptive threshold is 0.99.

In brief, we set two criteria for the transaction. The transaction is allowed when both conditions are satisfied at the same time. (1) $\sigma_t < \langle \sigma_t \rangle$ or $\left\langle \frac{dS_t}{S_t} \right\rangle > 0$, (2) $\frac{P_t}{\langle P_t \rangle} >$ threshold.

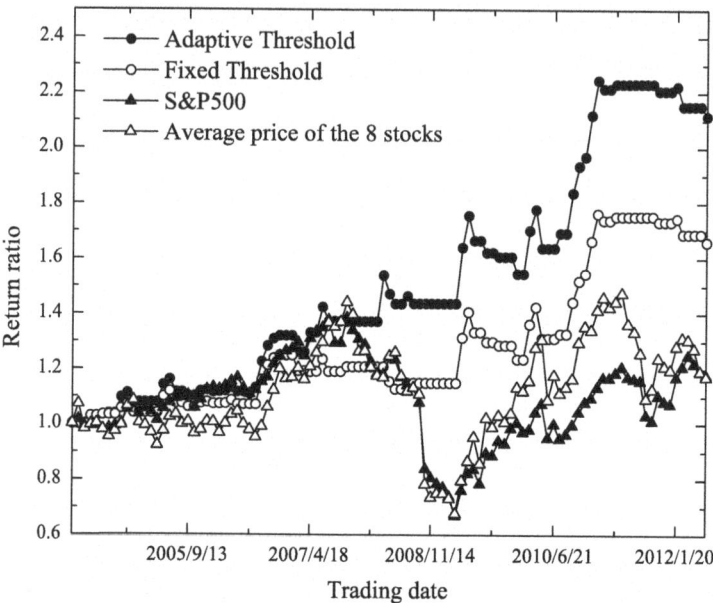

Fig. 1 The return ratio by the proposed trading strategy in comparison with the stock average price variation and the S&P500 Index over the same period

5 Simulation Result

To perform numerical test of our approach, we randomly select 8 stocks from S&P500 Index during the period between March 2004 and June 2012. We here choose (1) Applied Materials, (2) Boeing, (3) Cisco Systems, (4) Ford Motor, (5) IBM, (6) Juniper Networks, (7) Oracle, and (8) Xerox. In this work, the parameters are the trading period, sample size and they are 20 and 30 respectively. The fixed trading threshold is 1.01 and the adaptive trading thresholds are 0.99 and 1.01. The term return ratio, defined by the ratio of the final asset in cash value to the initial asset, is used as a measure for the performance of the strategy.

The results are shown in Fig. 1. We also include the S&P500 Index and the averaged stock price variation for comparison. The S&P500 Index shown is normalized by its initial value on the first trading days. The average stock price is defined by the arithmetic mean of daily closing price over the 8 stocks, and is normalized with respect to its initial value on the first trading day. At the end of the whole trading period, our strategy gives an increase of the asset by 111 % in adaptive threshold and 66 % in fixed threshold. Although the S&P500 Index also increases by 17 %, its results are somewhat mixed throughout the period when compared to the results of adaptive threshold and fixed threshold.

To compare the differences of these return ratios more quantitatively, we define the return ratio area which is a method to calculate the area differences

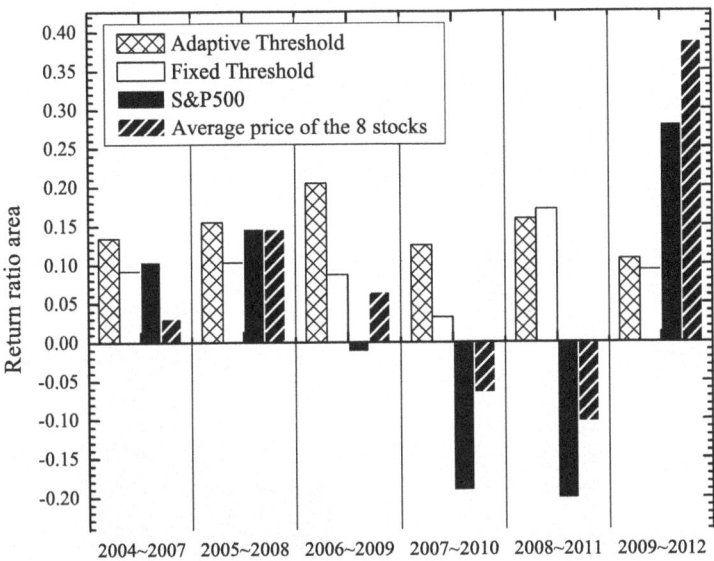

Fig. 2 The return ratio area by the proposed trading strategy in comparison with the stock average price and the S&P500 Index over the same period

of the return ratio. For example, if we obtain a return ratio function $f(x)$, we will subtract the area in the region of $f(x)$ where its value is below 1 from the area in the region of $f(x)$ where its value is above 1. If it is positive, then we will conclude that it is profitable and vice versa. We gather the trading results by calculating return ratio area for every 3 year period, and we have 6 return ratio areas from 2004 to 2012, namely, 2004–2007, 2005–2008, 2006–2009, 2007–2010, 2008–2011, 2009–2012 respectively. The return ratio areas of $f(x)$ are shown in Fig. 2. We again include the S&P500 Index and the averaged stock price for comparison.

One can see that the fluctuations of the return ratio areas of adaptive and fixed thresholds are less than that of S&P500 and averaged stock price. This implies that the proposed strategy can produce a portfolio of above-average return with less risk. During the period of the financial tsunami (2008–2009), the average stock price and the S&P500 index took a nosedive while our portfolio strategy is immune to the crash. The return ratio area shown in Fig. 2 during the period 2007–2010 also indicates our strategy can keep us from losing money in the financial tsunami and making profits during the rising period from 2009 through 2010. The sum of return ratio area which represents the average performance from 2004 to 2012 is shown in Fig. 3 which is the sum of the return ratio area shown in Fig. 2. The result from Fig. 3 suggests that the proposed trading strategy is better than the S&P500 Index and the average price of the 8 stocks.

We have also chosen other combinations of these 8 stocks and the average performance are shown in Table 1. From Table 1, it is clear that the method of

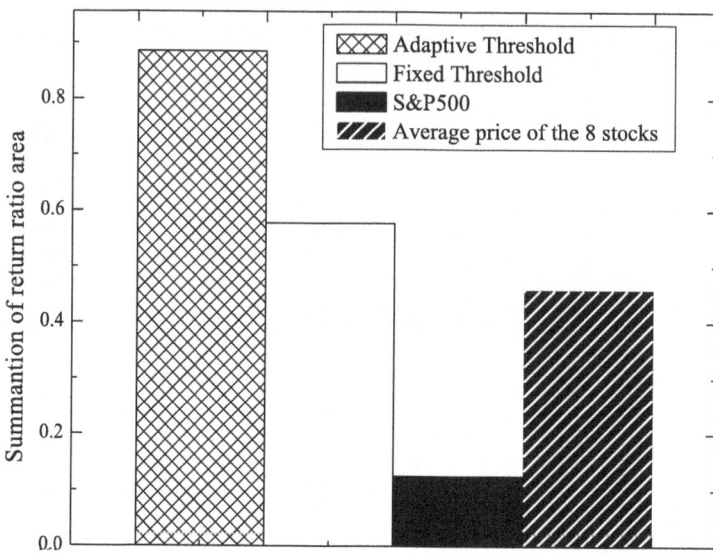

Fig. 3 The summation of return ratio area by the proposed trading strategy in comparison with the stock average price and the S&P500 Index over the same period

Table 1 The summation of return ratio area of 10 different 8-stock combinations in comparison with the stock average price and the S&P500 Index over the period from 2004 to 2012

Group	Adaptive threshold	Fixed threshold	S&P500	8 stocks average
1	0.5898	0.3454	0.1247	0.4514
2	0.4722	0.3094	0.1247	0.2327
3	0.5349	0.0890	0.1247	0.1828
4	0.4527	0.4123	0.1247	0.2097
5	0.3511	0.1177	0.1247	0.2240
6	0.3142	0.2705	0.1247	0.2790
7	0.6917	0.3037	0.1247	0.4489
8	0.5868	0.2021	0.1247	0.4059
9	0.4975	0.3490	0.1247	0.4778
10	0.7697	0.3616	0.1247	0.6475
Average	0.5261	0.2761	0.1247	0.3560

the adaptive threshold is the best strategy. We also notice that although the profit of fixed threshold is not always high, it seldom loses the trader's fortune during the whole period, in particular during the financial tsunami.

To further test the performance of our strategy, we follow Bacry's method [18] to generate an artificial stock index with the feature of multifractal random walk. Based on this artificial index, we generate 8 artificial stocks, each having 2,400 data points, meaning there are 2,400 trading days. Only the last 2,080 data points are chosen for trading, which means that the trading period is about 8 years (2004–2012, trading

Table 2 The summation of return ratio area by multifractal random walk in comparison with the stock average price and the S&P500 Index over the period from 2004 to 2012

Group	Adaptive threshold	Fixed threshold	Index	8 stocks average
1	0.01962	0.01177	0.01528	0.01715
2	0.02151	0.00285	0.02628	0.02547
3	0.00899	−0.00370	0.00547	0.00697
4	0.02517	0.01073	0.01429	0.01441
5	0.00721	−0.00005	0.00512	0.00494
6	0.01294	0.00523	0.01385	0.00976
7	0.00000	0.00397	−0.03423	−0.03354
8	0.02262	0.01054	0.02702	0.02740
9	0.01429	0.00451	0.00413	0.00332
10	0.01025	−0.00589	−0.01135	−0.01412
Average	0.01426	0.00399	0.00659	0.00618

days are about ∼260 × 8). The transaction results of the summation of return ratio area are shown in Table 2. It is clear that the results of the adaptive threshold are still better than the Index and the average price of the 8 stocks. The performance of our strategy also gives high return with low risk in this artificial stock market. Overall, our trading strategy can avoid crash, reflect the trend of market more sensitively and thus can make the transaction more favorable to traders and provide stable and consistent positive returns in the bullish as well as in the bearish markets.

6 Conclusion

In this paper, we have proposed a strategy of investment by time dependence variance analysis and index-trend prediction. The time dependence covers both the long term aspect of the pair of stocks, as well as the short term fluctuation of the stock prices. The long term aspect is determined by choosing the stock pairs with negative correlation at time t, while the short term stock price return provides a risk control through a trading threshold. By setting a suitable value for the trading threshold, we can avoid loss caused by a downward trend of the market. The overall performance of our strategy beats the performance of the average performance of the chosen set of stocks both in the simulated market and real market. This strategy should therefore be suitable for investors. Since mean variance analysis is a general method, the portfolio selected by this method should perform well in general for different stock markets. Moreover, since the stock purchase criteria incorporating the short term stock price fluctuations do not rely on the information of a specific stock market, we expect our analysis could be applied to any stock markets.

References

1. Markowitz H (1952) Portfolio selection. J Financ 7:77
2. Horasanl M, Fidan N (2007) Portfolio selection by using time varying covariance matrices. J Econ Soc Res 9(2):1
3. Hakansson NH (1971) Multi-period mean-variance analysis: toward a general theory of portfolio choice. J Financ 26:857
4. Maccheroni F (2009) Portfolio selection with monotone mean-variance preferences. Math Financ 19:487
5. Campbell J, Viceira L (2002) Strategic asset allocation-portfolio choice for long-term investors. Clarendon lectures in economics. Oxford University Press, Oxford
6. Schweizer M (1995) Variance optimal hedging in discrete time. Math Oper Res 20:1
7. Blanchet-Scalliet C, El Karoui N, Jeanblanc M, Martellini L (2008) Optimal investment decisions when time-horizon is uncertain. J Math Econ 44:1100
8. Fukunaga K (1990) Introduction to statistical pattern recognition, 2nd edn. Academic, Boston
9. Zemke S (1999) Nonlinear index prediction. Physica A269:177–183
10. Szeto KY, Cheung KH (1998) Multiple time series prediction using genetic algorithms optimizer. In: Proceedings of the international symposium on intelligent data engineering and learning, IDEAL'98, Hong Kong, pp 127–133
11. Szeto KY, Cheung KH (1997) Annealed genetic algorithm for multiple time series prediction. In: Proceedings of the world multiconference on systemic, cybernetics and informatics, Caracas, vol 3, pp 390–396
12. Szeto KY, Chan KO, Cheung KH (1997) Application of genetic algorithms in stock market prediction. In: Weigend AS, Abu-Mostafa Y, Refenes APN (eds) Proceedings of the fourth international conference on neural networks in the capital markets: progress in neural processing, decision technologies for financial engineering, NNCM-96, California. World Scientific, pp 95–103
13. Froehlinghaus T, Szeto KY (1996) Time series prediction with hierarchical raidal basis function. In: Proceedings of the international conference on neural information processing, ICONIP'96, Hong Kong, vol 2. Springer, pp 799–802
14. Szeto KY, Fong LY (2000) How adaptive agents in stock market perform in the presence of random news: a genetic algorithm approach. In: Leung KS et al (eds) IDEAL 2000, Hong Kong. Lecture notes in computer science/Lecture notes in artificial intelligence, vol 1983. Springer, Heidelberg, pp 505–510
15. Fong ALY, Szeto KY (2001) Rule extraction in short memory time series using genetic algorithms. Eur Phys J B 20:569–572
16. Chen C, Tang R, Szeto KY (2008) Optimized trading agents in a two-stock portfolio using mean-variance analysis. In: Proceedings of the 2008 Winter WEHIA & CIEF, Taipei, pp 61–67
17. Chen W, Szeto KY (2012) Mixed time scale strategy in portfolio management. Int Rev Financ Anal 23:35–40
18. Bacry E, Delour J, Muzy JF (2001) Multifractal random walk. Phys Rev E 64:026103

Self-Organization of Decentralized Markets with Network Externality

Xintong Li, Chao Wang, and Yougui Wang

Abstract In this paper an agent-based model is developed to simulate the evolution of a decentralized market with network externality. The traders in the market who are characterized by their willingness prices adjust their ask or bid prices to maximize their own individual surplus subject to the social effect of cumulative transaction. The decentralized market eventually exhibits not one single equilibrium but a stable state with obvious price dispersion. It is found that self-organization of the decentralized market is path dependent and locked in local rather than global maximum. Network externality will enhance or mitigate the expansion of the transaction driven by self-optimization of both sellers and buyers, thus having significant impacts on the final trading volume and market efficiency at stable states.

1 Introduction

During the past 20 years, agent-based models used to analyze the market processes [5, 7], which is interpreted as a collection of agents interacting with each other [12], have proven to be an effective bottom up modeling method [11] to simulate market dynamics on the micro scale. Researchers endow their artificial agents with certain preferences and set simple, adaptive learning rules on the way they perceive in modeling exercises. As these artificial agents interact with one another and their environment, adaptation takes place via relative fitness considerations at both the individual level as well as the collective level. All these details of how artificial agents adapt are less important than the aggregation outcomes that emerge from repeated interactions. These analytical methods are much more preferred especially when the emergent outcomes cannot be deduced without resorting to

---

X. Li · C. Wang · Y. Wang (✉)
Department of Systems Science, School of Management, Beijing Normal University, Beijing 100875, People's Republic of China
e-mail: jnlixt@163.com; wangchao_26@126.com; ygwang@bnu.edu.cn

simulation exercises. Neo-classical economics focuses on the descriptions of the transition of markets from an equilibrium state to another one. In most cases, there is a lack of understanding on non-equilibrium state of a market and its dynamics. Individuals' behavior and their interaction provide us with a brand new perspective from which we can have a better understanding of the market. Different from the general equilibrium analysis, the interaction among individuals on the micro scale results in macro phenomenon including the emergence of equilibrium prices.

Albeit with this approach, whether the efficient outcome of markets can be realized by self-organization of the traders is still unsolved so far. Standard economic theory is built on two specific assumptions: utility-maximizing behavior and the institution of Walrasian tatonnement [3]. A powerful coordinator must exist who possesses all relevant information for transactions and governs a Walrasian tatonnement. Smith argued that the market can still get efficient with Walrasian tatonnement replaced by double auction [9]. Gode and Sounder further proved this conclusion in a generalized way [2,3] later. They examined the behavior of markets with zero-intelligence traders who submit random bids and offers. In contrast to the evolutionary models of [8], natural selection plays no role in arriving at Gode and Sounder's conclusions; the surplus extraction property of double auctions is attained by an unchanging population of zero-intelligence traders. The primary cause of the high allocation efficiency of double auctions is the market discipline imposed on traders while learning, intelligence, or profit motivation are no longer necessary. When such market mechanisms are embodied, aggregate rationality may be generated even from individual irrationality.

The performance of an economy is the joint result of its institutional structure, market environment, and agent behavior. Institutional structure is defined by the rules that govern exchange, market environment by agents' tastes and endowments of information and resources, and agents' behaviors by their trading strategies (see [10]). The absence of rationality and motivation in the zero-intelligence traders seems to be offset by the structure of the auction market through which they trade. However, the double auction still needs a center in essence playing the same role as the market maker in Walrasian tatonnement to gather the ask-bid information and make the decision so that the equilibrium transaction requirements can be fulfilled. To sum up, all these centralized disciplines that play an important role in the convergence of the transaction prices to equilibrium levels call for an information processor as a powerful coordinator who collects and sequences the bargaining prices constantly submitted to the market.

No one is an island. Even when no public information is explicitly provided, and all agents can only make judges and incremental adjustment based on personal experiences of his or her own surplus varying over time, we can still expect that these agents are receptive to the overall situation of the market. Writing in 1950, Harvey Leibenstein analyzed the "bandwagon effect", by which he meant the extent to which the demand for a commodity is increased due to the fact that others are also consuming the same commodity. It represents the desire of people to purchase a commodity in order to get into 'the swim of things'; in order to be fashionable or stylish; or, in order to appear to be 'one of the boys' [6]. However, that is not always the case. The strong longing for original style and personalized charm drives people to pursue unique and distinctive consumption instead of simply drifting with the current. These conditions fall into two categories of the manifestation of the social connection among agents in the form of network externality, which has been defined as a change in the benefit, or surplus, that an agent derives from a good when the number of other agents consuming the same kind of good changes.

For supporting examples, one needs to look no further than a fax machine. As fax machines become increasingly popular, for example, an increase in the value of one's fax machine can be anticipated since he or she will have much use for it. This allows, in principle, the value received by consumers to be separated into two distinct parts. One component, usually labeled as the autarky value, is generated by the product even if there are no other users. The second component, which has been called as synchronization value, represents the additional value derived from being able to interact with other users of the product. It is the latter that captures the essence of network externality.

In this paper we attempt to see whether a market without a market maker or 'decision-making centre' can achieve its efficiency by observing the market evolution with adaptive and interactive traders. With the acceptance of network externality as the buyer set expands, the decentralized market will be portrayed not as simple but complex, not as deterministic, predictable and mechanistic, but instead as process-dependent, organic and always evolving [1].

2 Model

Assuming that there is only one kind of good to be traded in a market, here we put forward a price-based self-optimization model to demonstrate the market in non-equilibrium conditions. The whole system consisting of N sellers and N buyers evolves in a trial-and-error dynamic process, during which the supply, demand, and price vary in a prescribed way. The traders are a group of adaptive learners who are assigned to seek for their best ask or bid price. On their entry into the market, the sellers and buyers have their initial willingness prices based on their personal valuations of the given good. For example, the willingness prices for the j_{th} seller and the i_{th} buyer are denoted by V^{sell}_{0j} and V^{buy}_{0i} respectively. The values of V^{sell}_{0j} or V^{buy}_{0i} are uniformly distributed over a closed interval $[a, b]$.

The original bargaining prices are randomly chosen in the valid interval in accordance with individual initial willingness to exchange during the first two periods. A period is defined as R rounds (usually $R = N$) of randomly paired-up trading. We note that the sellers and buyers adopt a consistent strategy, i.e., they hold constant bargaining prices within one period and update it according to the surplus calculation between the periods. In any one period, each buyer will repeat the process of encountering one seller randomly to make transaction for R times. For instance, at the r_{th} round in the t_{th} period, suppose that the i_{th} buyer and the j_{th} seller meet with each other and their bargaining prices for the current period are $P^{Ask}_{t\,j}$ and $P^{Bid}_{t\,i}$ respectively, if and only if $P^{Ask}_{t\,j} \leq P^{Bid}_{t\,i}$, the exchange between the two traders will be realized. And we have the price for the deal, incremental surplus for the buyer and the seller as follows,

$$Price_{t\,ij} = \frac{P^{Ask}_{t\,j} + P^{Bid}_{t\,i}}{2},$$

$$B_{t,r_i} = V^{buy}_{t\,i} - Price_{t\,ij},$$

$$\Pi_{t,r_j} = Price_{t\,ij} - V^{sell}_{t\,j}.$$

Otherwise, i.e., $P^{Ask}_{t\,j} > P^{Bid}_{t\,i}$, we have

$$B_{t,r_i} = 0,$$

$$\Pi_{t,r_j} = 0.$$

Thus the total surplus that buyer i and seller j would obtain in this period are given respectively by

$$B_{t\,i} = \sum_{r=1}^{R} B_{t,r_i}$$

$$\Pi_{t\,j} = \sum_{r=1}^{R} \Pi_{t,r_j}.$$

The amount of transaction in the t_{th} period is denoted by Q_t. During the first two periods, each buyer will develop his expectation of the quantity of transaction Q_e on the basis of Q_1 and Q_2, taking mathematical average of the two as

$$Q_e = \frac{Q_1 + Q_2}{2}.$$

For simplicity, the step of the incremental adjustment δ is positive, preset and identical for all agents, which is relatively small compared with the bargaining prices of both buyers and sellers. The adaptive behaviors of the agents are described by adjusting their ask prices and bid prices step by step to be better-off. For each adjustment, one has a binary choice to make: to raise or lower his price for the following period of trading. All agents simultaneously decide their prices given information about previous transactions concerning his own surplus. Therefore the mathematical formula of price adjustment for the i_{th} buyer and the j_{th} seller can be written respectively as

$$P^{Ask}_{t+1\,j} = P^{Ask}_{t\,j} + sign((\Pi_{t\,j} - \Pi_{t-1\,j})(P^{Ask}_{t\,j} - P^{Ask}_{t-1\,j}))\delta,$$

$$P^{Bid}_{t+1\,i} = P^{Bid}_{t\,i} + sign((B_{t\,i} - B_{t-1\,i})(P^{Bid}_{t\,i} - P^{Bid}_{t-1\,i}))\delta.$$

To introduce network externality to our analysis, we assume that each buyer's willingness to pay for the good in current period depends not only on his original preference and valuation but also on the number of other buyers who have purchased the good. Thus the willingness price of the i_{th} buyer in period t can be expressed as follows,

$$V^{buy}_{t\,i} = V^{buy}_{0\,i} + W_{t-1\,i},$$

$$W_{t\,i} = \alpha \frac{(Q_t - Q_e)}{Q_e} V^{buy}_{0\,i}.$$

Meanwhile the sellers all keep their initial willingness prices unchanged forever after entering the market. $W_{t-1\,i}$ denotes the increment of willingness price of the i_{th} buyer due to one's awareness, which is acquired in the $(t-1)_{th}$ period, of the potential social addition or damage associated with a purchase. The quantity of the realized transaction beyond expectation urges the buyers to increase or decrease their valuation of the utility brought by the purchase. As a consequence, the buyers vary their willingness prices accordingly. To be specific, in the expression above, α is the parametric representation of the intensity and direction of the social interaction effect on the market as a whole, whose sign indicates the direction of immediate market feedback. A plus one represents the desire of people to purchase a commodity in order to conform with the people they wish to be associated with. A minus one, on the contrary, represents that the social interaction curbs the transaction against the incentive of self-interest which is expected to increase the trading volume. Since those with higher initial willingness prices tend to be more sensitive to social trend and more prone to be influenced by others, $W_{t-1\,i}$ is in direct proportion to $V^{buy}_{0\,i}$ in our model to indicate the heterogeneous individual susceptibility of the buyers to the network externality.

3 Results and Discussion

There are sufficient sellers and buyers in the competitive market who will transact equally since not only can the sellers set their ask prices but also the buyers offer their bid prices in a transaction. In other words, both sellers and buyers have the same power in this evolving market. We simulate such price searching for 1,000 periods ($T = 1,000$) of 1,000 buyers and 1,000 sellers ($N = 1,000$) each endowed with a willingness price randomly chosen from $[0, 100]$. Since Q_e is set as the trading volume formed on the entry of all the agents into the market, the social effect of cumulative transaction will take effect gradually instead of causing fallacious disturbance from the start. The incremental step of adjustment is designated to be $\delta = 1$. With all things being equal, we choose a series of α ranging from -1.0 to 1.0 to investigate the effect of the network externality and record the actual quantity of transaction and bargaining prices of all traders in each period ($t = 1, 2, 3, \ldots, 1,000$).

As shown in Fig. 1, the evolutions of the quantity of transaction for all cases exhibit the same profile: starting with a low level of trading volume, they increase gradually and attain a final state of definite value respectively after sufficient rounds of simulations. Despite the effect of network externality, the trading volume will instinctively be driven up by self-optimization of both sellers and buyers. In comparison with such a result, positive network externality will enhance the growth of the transaction, leading the trading volume at final stable state to a higher level strikingly. Negative network externality, on the other hand, mitigates such progress by partially offsetting, not eliminating, the expansion of transaction, thus yielding a lower final trading volume.

Mathematically, given the parameter α, the quantity of transaction at the stable state of the market can be deduced by the supply and demand curves in period t that can be formulated by the cumulative distribution of the current bargaining prices of the sellers and buyers respectively [13]. Compared to theoretical equilibrium in a market of Walrasian tatonnement or double auction with the same initial state where the willingness prices of both sellers and buyers are uniformly distributed on $[0, 100]$, the decentralized market composed of N ($N = 1,000$) agents is in fact inefficient through the mechanism of self-organization. It can be reflected from the twisted supply and demand curve whose intersection deviates from $(\frac{N}{2}, \frac{a+b}{2})$, as shown in Fig. 2. This can also be reflected from the realized proportion of optimal social welfare. Following Smith [9], one can define the efficiency of markets as the total profit actually earned by all the traders divided by the maximum total profit that could have been earned by all the traders. From what formulates the initial state of the market, the market's demand and supply curve will cross at $(\frac{N}{2}, \frac{a+b}{2})$ as the competitive equilibrium if all individuals are still guided by the collective information to the efficient state as is in Walrasian tatonnement or double auction. In this case, the maximum total profit E_0 can be derived from the sum of the seller and buyer surplus as follows,

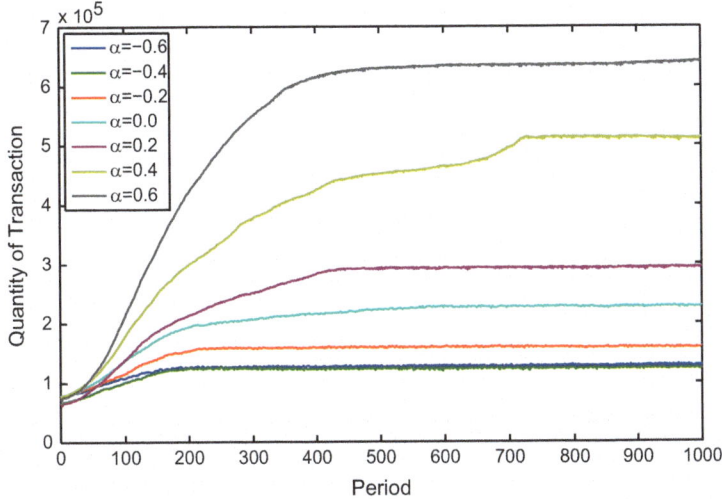

Fig. 1 The evolutions of the quantity of transaction in the markets with different network externality

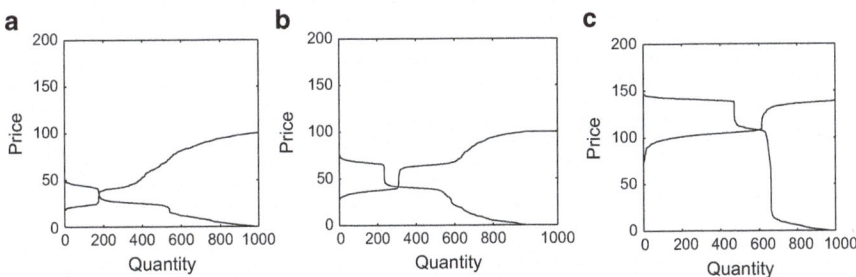

Fig. 2 The deduced supply and demand curve of the market at final stable state for three cases of (**a**) $\alpha = -0.5$, (**b**) $\alpha = 0$, and (**c**) $\alpha = 0.5$

$$E_0 = \sum_{i=1}^{N}(V^{sell}_{0i} - V^{buy}_{0i}),$$

when there is $V^{sell}_{0m} \leq V^{sell}_{0n}$ for any $m \geq k$, and $V^{buy}_{0l} \geq V^{buy}_{0k}$ for any $l \geq k$. The total surplus divided by E_0 can thus function as quantifiable measurements of the realization of the welfare of all concerned, namely the period-by-period efficiency of the markets governed by decentralized self-organization, and is shown in Fig. 3. Based on the preceding analysis, we can exhibit the relationship between α and the final quantity of transaction together with that between α and the final efficiency at the corresponding stable states in Fig. 4.

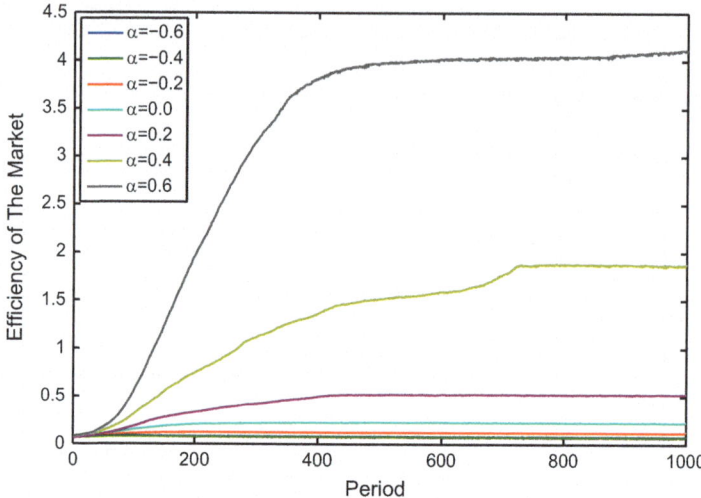

Fig. 3 The evolutions of efficiency of transaction in the markets with different network externality by the standards of E_0

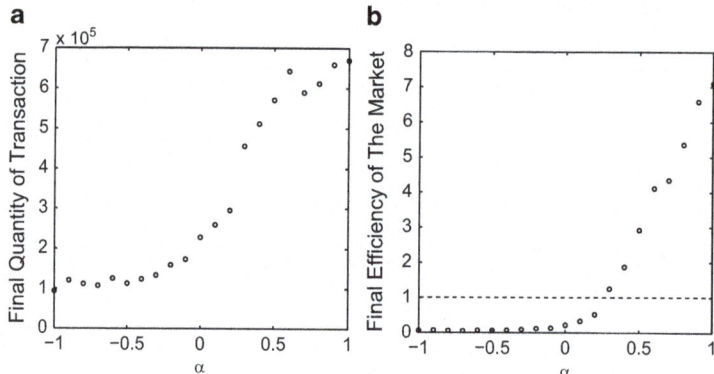

Fig. 4 The quantity of transaction and market efficiency in the markets with different network externality at final stable state

In order to separate the influence exerted by the social effects from the evolving system, we evaluate the efficiency of the decentralized market with network externality by referring to the total surplus gained at equilibrium of this market yet free of any possible social effects as a standard. Set α to be zero for simulation, the ultimate stable state gives its surplus as the denominator E_{nex}. The total surplus divided by E_{nex} can similarly function as quantifiable measurements of the realization of the welfare of all concerned and is shown in Fig. 5.

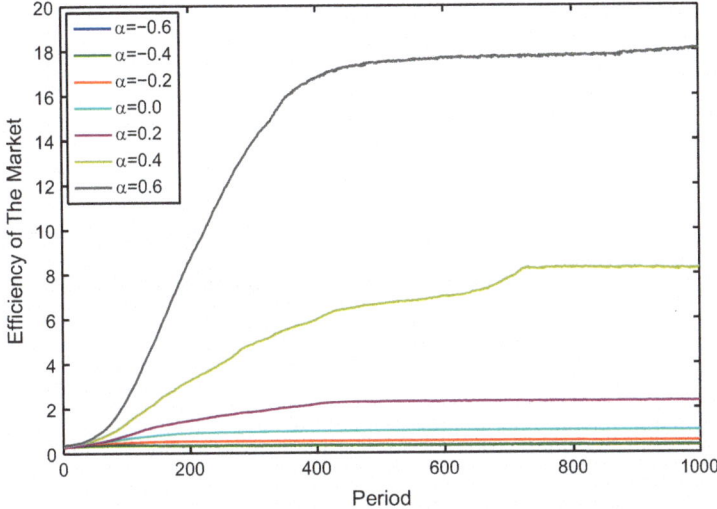

Fig. 5 The evolutions of efficiency of transaction in the markets with different network externality by the standards of E_{nex}

In our simulations, with the buyers and sellers interacting with each other the market does not evolve to a single equilibrium as predicted by the classical competitive market model but to a state of price dispersion in regard to their original bargaining prices depicted in Fig. 6 and to their original willingness prices in Fig. 7. As the transactions proceed, we can observe that the ask-bid prices of all agents vary over time. The bargaining prices of buyers and sellers originally distributed below and above the diagonal line will eventually form three separate layers respectively at the stable state. It can also be readily concluded that the positive network effect drives the bargaining prices up and results in more distinct layers of both ask and bid prices, which implies a higher market efficiency. The negative network effects, on the contrary, generate stable states with the formation of less obvious aggregation due to the lack of incentive for the buyers to set high valuation on the purchases which finally lowers the efficiency of the market. At this point, not only can we interpret the stair-step deduced demand and supply curve from the perspective of dispersed bargaining prices, but also two plausible explanations present themselves which account for the enhancement of positive network externality to market efficiency. Firstly, it is self-evident that the perception and reception of social addition by itself will increase the surplus of the buyers in the market. Secondly, the distinct differentiation caused by positive network externality naturally results in better convergence of the aggregation and improved matching between two sides of the market.

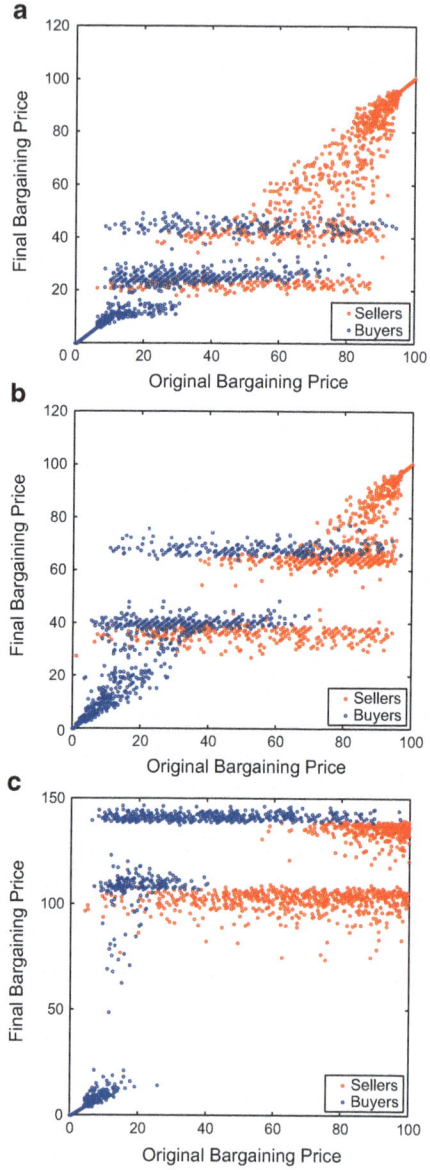

Fig. 6 The final bargaining price versus the original bargaining price for three cases of (**a**) $\alpha = -0.5$, (**b**) $\alpha = 0$, and (**c**) $\alpha = 0.5$

It is also enticing to figure out what generates such a perplexing phenomenon. The local incremental adjustment mechanism that differs from the one of the Walrasian tatonnement or double auction together with the path dependent characteristic of the market evolution are both essential to the formation of price dispersion. The market is constantly changing for all the agents, which means

Fig. 7 The final bargaining price versus the original willingness price for three cases of (**a**) $\alpha = -0.5$, (**b**) $\alpha = 0$, and (**c**) $\alpha = 0.5$

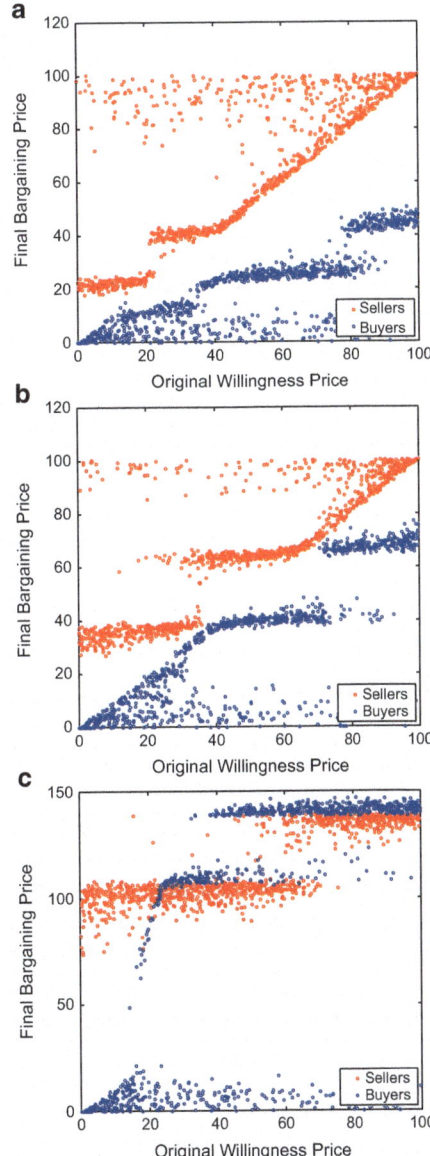

that each seller encounters an altering demand curve, whose optimized price in the previous period may no longer be a good one when his ask price moves to that point. The stable state is not deduced by the agents locating at their globally maximizing ask-bid prices. This path dependent process can be anticipated in a self-organized system with no information concentration center. No matter whether in a market of Walrasian tatonnement or double auction, individuals are all guided by the collective information to get to the market efficient state. From this aspect, it is not the institution itself but the collective information revealed by it makes the market efficient. Therefore the main function of a market institution is to reveal as much collective information as possible and that is how market economy arrives at efficiency.

When the system is predetermined without the randomness of matching between sellers and buyers, the evolution theoretically leads to the conclusion that similar pattern of price dispersion yet with more layers will appear at the stable state. As shown in Fig. 8, when the agents' surplus for each period is calculated by traversing the complete set of their counterparts, there is little possibility to avoid local maximum of the system. At the same time, greater volatility is observed in the fluctuation of bargaining prices of every single agent as well. Compared to the pervious cases, it is the introduction of random matching into the system that induces more agents to the same price level as the equilibrium price in competitive market. However, it is also expected that excessive uncertainty brought by the stochastic process will result in an utterly chaotic final state of the system. It reveals the dual character of the randomness functioning in the market as both data miner and detrimental noise.

4 Conclusion

We are trying to answer the question whether the efficient outcome of a market can be realized by the self-organization of all traders. We have already known that a market with an institution of Walrasian tatonnement and double auction will get close to the competitive market equilibrium even with zero-intelligence traders [4]. But all these market institutions have a function of gathering information from micro level to macro level. It is not surprising that a centralized market reaches an efficient state.

In our model of a decentralized market with network externality, the buyers and sellers make transactions randomly and adjust their ask-bid prices adaptively. The market evolution is driven by the self-organization of all the agents without manipulation from any information center. The result shows two typical effects of this market mechanism. Firstly, the market will evolve to a stable state with price dispersion. This phenomenon can be explained by the fact that the evolution of traders' ask-bid price is path dependent and a portion of the sellers and buyers are trapped at local maximum point. Secondly, there is a scope of making gains under positive externality and suffering losses under negative externality in market efficiency. This result suggests that network externality should not be neglected in discussions of the efficiency of a decentralized market.

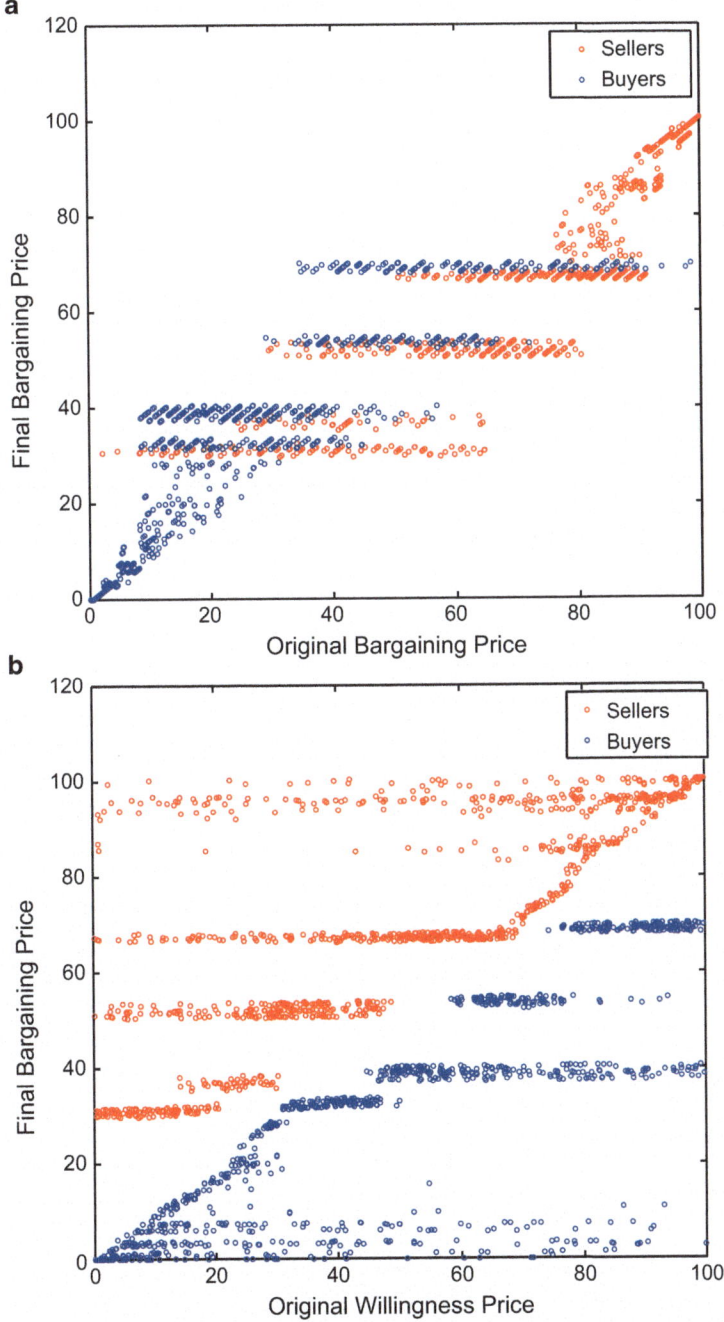

Fig. 8 The final bargaining price versus (**a**) the original bargaining price, and (**b**) the original willingness price when $\alpha = 0.0$ without random matching

Acknowledgements We thank Dr. Jianzhong Zhang for his helpful comments and language improvements. This research was supported by National Natural Science Foundation of China under Grant of No. 61174165 and Program for New Century Excellent Talents in University (NCET-10-0245). This work was also the result of Interdisciplinary Salon of Beijing Normal University.

References

1. Arthur WB (1990) Positive feedbacks in the economy. Sci Am 262:92–99
2. Bosch-Domenech A, Sunder S (2000) Tracking the invisible hand: convergence of double auctions to competitive equilibrium. Comput Econ 16:257–284
3. Gode DK, Sunder S (1993) Allocative efficiency of markets with zero-intelligence traders: market as a partial substitute for individual rationality. J Politi Econ 101:119–137
4. Gode DK, Sunder S (1997) What makes markets allocationaly efficient? Q J Econ 112(2): 603–630
5. Krawiecki A, Holyst JA, Helbing D (2002) Volatility clustering and scaling for financial time series due to attractor bubbling. Phys Rev Lett 89:158701
6. Leibenstein H (1950) Bandwagon, Snob, and Veblen effects in the theory of consumers demand. Q J Econ 64:183–207
7. Lux T, Marchesi M (1999) Scaling and criticality in a stochastic multi-agent model of a financial market. Nature 397:498–500
8. Nelson RR, Winter SG (1982) An evolutionary theory of economic change. The Belknap Press of Harvard University Press, Cambridge MA
9. Smith VL (1962) An experimental study of competitive market behavior. J Politi Econ 70: 111–137
10. Smith VL (1982) Microeconomic systems as an experimental science. Am Econ Rev 72: 923–955
11. Tesfatsion L (2001) Special issue on agent-based computational economics. J Econ Dyn Control 25:281–654
12. Tesfatsion L (2003) Agent-based computational economics: modeling economies as complex adaptive systems. Inf Sci 149:262–268
13. Wang Y, Stanley HE (2009) Statistical approach to partial equilibrium analysis. Phys A 388:1173–1180

Who Wins: Yoda or Sith? A Proof that Financial Markets Are Seldom Efficient

Lucian Daniel Stanciu-Viziteu

Abstract We propose an artificial financial market where three types of investors compete. Value investors, that use information to align the asset's price with it's value are called YODA.[1] SITH is our name for the investors who hold information but decide not to use it right away, and instead act as non-informed investors. All other agents trade without information. We show that SITH agents can make better risk-adjusted gains than YODA agents. Consequently we prove that informed investors have incentives to withhold information and act like chartist traders. Our observations lead us to state that financial markets are consistently overpricing assets and can be regarded as *seldom* efficient.

1 Introduction

The seminal work of [5] proves that financial markets cannot be informationally efficient. When information is readily available in market prices investors lose interest in acquiring new information. As a consequence, new market prices become deprived of fundamental information which makes information once again profitable. After this proof, methods for financial research have evolved and were transformed by multi-agent models [2] that create frameworks for explaining real market price dynamics and strategy competition. As late as 1980, in works like [3] and [4], different investment strategies have competed to yield market realistic scenarios (e.g. stylized facts). Some results show no clear strategy dominance but rather a

---

[1] YODA/SITH are the main positive/negative characters of the famous movie series "Star Wars" created by George Lucas.

L.D. Stanciu-Viziteu (✉)
CERAG UMR CNRS 5820, University of Grenoble, Grenoble, France
e-mail: lucidinbr@gmail.com

continuous change in strategy distribution. With this article we contribute to the existing literature on investing strategy competition [7, 11] by proposing a model where we have a proxy measure for efficiency (the distance between price and fundamental value) and show that withholding information can be profitable. Our results validate the Grossman and Stiglitz paradox and add to it by showing that prices are consistently inflated. To arrive at these results we design a multi-agent financial market simulator. Using this simulator we show that informed investors make the best risk-adjusted returns if they decide, in certain conditions, to trade like Chartist investors. Our market model provides a parsimonious way to: (a) create an information flow from an unobservable theoretical fundamental value, (b) endow investors with comprehensive behavioral traits and (c) build a simple measure of market efficiency. Our goal is to discover when market inefficiency appears consistently.

2 The Model

In this study we propose a model of a financial market with a single stock with market price P_t. We suppose this stock has a fundamental value series V_t, with normally distributed returns and[2] $V_0 = P_0$, and this value is not known directly by agents. The fundamental value should be considered as a theoretical construction and not as a series that exists in real life. We assume that this value is not observable yet investors, using different stock analysis methods, can infer certain estimators.

$$V_t = V_{t-1} * e^{N(\mu V, \sigma V)} \tag{1}$$

Information is a function, see Eq. 2, of the distance between the market price and the fundamental value, biased by investor specific characteristics. If the market price is above the fundamental value then information is negative, representing an inflated price. When the market price is below the fundamental value the information has a positive sign and signals an underestimation of the stock's value. Acknowledging the advances in behavioral finance, see [6], we allow for heterogeneity among the investors' interpretations of the information. The parameters a_x and b_x are specific to each investor and they represent the manner in which information is interpreted. Different combinations of these parameters yield specific investor behaviors, as described in Table 1.

$$I_t^x = a_x * (V_{t+1} - P_t) + b_x \tag{2}$$

As Eq. 2 shows, our information stream is measured in the same units as the stock price. Investors do not have to discount this information since it is already

---

[2]This first price can be considered as an exogenous IPO price.

Table 1 Combinations of A and B values and their respective behaviors

$a = 1$	$b = 0$	Perfectly informed investor
$a > 1$	$\forall b$	Exaggerating investor
$0 < a < 1$	$\forall b$	Conservative investor
$\forall a$	$b > 0$	Optimistic investors
$\forall a$	$b < 0$	Pessimistic investors

discounted. This method is intended to simplify the market model and allows us to easily observe the relations between market price, information and fundamental value. The fact that the information received by investors is discounted does not imply that all investors construct future price expectations using the same discount factors.

The trading period t spans from time t to $t + 1$. At the beginning of trading period t the fundamental value of the asset changes, instantaneously, from V_t to V_{t+1} and remains constant until the end of the period. Investors receive information I_t which informs on how has the asset evolved since the last market price. Investors use this flow of information in their search for profits. The new market price P_{t+1} reflects the traders expectations of the new fundamental value V_{t+1}. In a market with rational expectations [10] the market price should follow Eq. 3, where ϵ has a zero mean.

$$P_t = V_t + \epsilon \qquad (3)$$

Our simulated market is populated by informed and uninformed investors. We consider two types of uninformed investors: Noise and Chartist strategies. A Noise strategy mimics the behavior of liquidity investors who only use stocks for very short-term investments, disregarding information. Because Noise traders do not care about prices they only send market orders (without any price quote attached). Therefore, these traders do not have any direct effect on the formation of market prices. Their role is to supply liquidity to the market. The other type of uninformed investors are Chartists, which care for market prices and only send limit orders. Chartist investors use simple technical forecast rules, like $\{-1,0,1\}$, for buying and selling stocks. A rule is activated when the historical market returns are consistent with the rule specification. The previous rule is activated when $r_{t-2} < 0$, $r_t > 0$ and we are indifferent of r_{t-1}. Chartists are endowed, at the beginning of each simulation, with a selling and a buying rule. When a rule is activated, Chartists will try to sell/buy as much as possible with a limit order price as specified in Eq. 4.

$$P_{order} = P_t \pm N(\mu_c, \sigma_c) \qquad (4)$$

The parameters μ_c and σ_c represent the Chartist's aggressiveness in terms of trend following. A high μ_c implies that a Chartist will expect high yields from the market and will bid accordingly. At the beginning of a simulation each Chartist is endowed with a combination of values of μ_c and σ_c, which remain constant.

Our market model distinguishes two main types of informed investors: YODA and SITH. Both investor types have access to the same information flow yet their views about the market are different. YODA investors believe markets are, in general, efficient. They acknowledge that there may be situations, due to inconsistent information or lack of liquidity, when market prices can temporarily stray from fundamental values. Therefore YODA investors will always trade according to the information they receive. When prices are too high (negative information) they sell and when prices are too low they will buy. This strategy is sometimes referred to as a *fundamentalist* strategy.

On the other hand SITH investors believe prices can be disconnected from their fundamental value for sustained periods of time. Therefore, SITH investors will not always use their information in a 'value-restoring' fashion. More exactly, if the information shows an overappreciation of the stock the SITH investors can still buy the asset. Yet, when the information shows a too greater distance between price and fundamental values the SITH investors will act like YODA investors. This change in strategy is due to a risk tolerance parameter Y specific to the SITH investors. This risk tolerance level is specific to each investor.

Regardless of their views about the market, all informed investors are risk averse. To model this aversion we equip each informed investor with a parameter called minimal return expectations r_{min}. This parameter represents the combination of three important factors: the risk-free rate, the investor's assessment of the asset risk and his risk premium. An informed investor decides to enter a trade only if the expected return from the trade is higher or equal to his minimal return expectations. During the trading period t, stretching from time t to t+1, informed investor x computes the expected value of the future market price as in Eq. 5.

$$E_t^x(P_{t+1}) = P_t + I_t^x \tag{5}$$

Because informed investors have positive return expectations they will trade only if their expected price estimation is far enough from the last market price. Therefore, each informed investor has a confidence interval for his expected market price within which he does not trade, since it would not yield enough returns. This confidence interval is computed as

$$IC_t^x(P_{t+1}) = [E_t^x(P_{t+1}) * (1 - r_{x,min}); E_t^x(P_{t+1}) * (1 + r_{x,min})] \tag{6}$$

The confidence interval $IC_t^x(P_{t+1})$ is interpreted by investor x as follows: *Given the market evaluation of the stock, my interpretation of the recent available information, the risk free rate, my risk premium and risk assessment about this stock than the next period market price of the stock should be inside the interval $IC_t^x(P_{t+1})$. Therefore, if this period's market price P_t is outside this interval it is worth trading since the expected value of the trade is positive.*

If $P_t < Min(IC_t^x(P_{t+1}))$ then buy at the maximum price of $Min(IC_t^x(P_{t+1}))$
$$\tag{7}$$

If $P_t > Max(IC_t^x(P_{t+1}))$ then sell at the minimum price of $Max(IC_t^x(P_{t+1}))$
$$\tag{8}$$

If the last market price is above/below the confidence interval of the investor's expected one-period-ahead market price estimation then the investor will send a limit buy/sell order for a maximum/minimum price of $Min(IC_t^x(P_{t+1}))$, respectively $Max(IC_t^x(P_{t+1}))$. In this way YODA informed investors attempt to cash in on the current market mispricing that, they hope, will be resolved in future trading rounds. This type of investor strategy tends to restore market efficiency.

On the other hand, SITH investors use a mixed strategy. They estimate the future market price and it's confidence interval in the same way as YODA investors. The difference is that, while the current market price is not to far away from their expected confidence interval, SITH investors will not rest passive but adopt a Chartist strategy to profit from trend building. Each SITH investor x has a risk-limiting parameter, Y_x, which represents a maximum distance between market price and fundamental value that the SITH investor accepts as being "normal" and until which he will use a Chartist strategy.

$$\text{If } Y_x > \frac{E_t^x(P_{t+1})}{P_t} \text{ then use YODA strategy} \qquad (9)$$

When the market price is too far from the future price expectations, computed using the biased flow of information, a SITH investor will stop using a Chartist strategy and adopt a YODA type strategy (betting on the reversal of price towards it's fundamental value). From an aggregated point of view, the SITH investors help fuel price trends and also profit off of the reversal, while YODA investors only try to revert prices back to their fundamental level.

Non-informed investors do not care for the market price and they use only market buy/sell orders. These orders are resolved at the new market price. All other investors, chartists, SITH and YODA, send limit orders to buy or sell stock at a maximum/minimum price. These orders are collected in a limit-order book that computes the new market price in a four step procedure[3]: (a) maximize executable volume (b) establish minimum surplus (c) dampen market pressure (d) use last market prices as reference for the new price. This market price discovery procedure is similar to stock market opening price fixing and provides our model with a extra-day, non-continuous, trading scheme. This method assures, as proved in [9] that market pressure and information are correctly and fully incorporated in the resulting market price. Reference [8] considers that the limit-order book price formation mechanism is the best method to be used in the design of multi-agent financial market simulators.

When sending orders, our modeled investors try to buy stocks for all of their cash or sell their entire portfolios. This choice of quantity can be argued in two ways. Firstly, we consider that holding cash (or the risk-free asset) does not provide any returns thus a positive expected risky return is enough motivation for investors to hold only the risky asset. Secondly, investors enter the market only when their

---

[3] As used by the Australian Stock Exchange and explained in http://www.asx.com.au/products/calculate-open-close-prices.htm

expected returns are greater than their minimal expected returns, which combines the risk-free return and the asset's risk premium. Introducing a separate risk-free asset with a positive return would not impact our results but would only make the model more complex (especially in the construction of the investor's portfolio). The next section lists some of the interesting results we obtain using this multi-agent model of a financial market.

3 Results

To justify the title statement and our research goal we prove three hypothesis:

1. Chartists traders, not necessarily in overwhelming numbers, consistently move market prices away from fundamentals. These traders can survive in markets for extended periods of time.
2. A mixed fundamental-chartist strategy is more profitable than a pure fundamental strategy even on a risk-adjusted basis.
3. A chartist or a mixed fundamental-chartist strategy can survive in financial markets even in the absence of non-informed traders. Therefore, informed investors have incentives to trade using non-informed strategies (a behavior that increases market inefficiency).

If Chartists consistently create price bubbles the market price will be, on average, upward biased. Moreover, these upward biases are not predictable. It is arguable that chartists can create reverse price bubbles, like the 'fire-sales' described in [1], but the intensity of these bubbles is very weak. Diminishing market prices enable informed investors to buy more assets at a discount price thus dampening the bubble. To test the first hypothesis we simulate a financial market where perfectly informed investors compete against chartist investors. We are interested in determining when can the chartist investors consistently create price bubbles and for how long they can survive in the market. For each test we present a table with the most important parameters that shape the simulated market. The first simulation is governed by the parameters from Table 2. All of our simulation results exhibit the *stylized facts* proprieties.

We vary parameter x and observe the statistical proprieties of the market price bubbles. The size of the market mispricing is measured using the distance between the market price P_t and the corresponding fundamental value V_t. All the numbers reported in the table are statistically significant (for each level of x we run multiple tests and report the average statistics).

We observe, in Table 3, that price bubbles appear when chartist traders represent at least 35 % of the traders' mass. A sample price evolution of a market with 35 % chartist and 65 % perfectly informed traders can be seen in Fig. 1. When this critical mass of non-fundamental investors is reached the market price departs often from it's fundamental value. Therefore, the assumption stated in Eq. 1 is violated because the error term has a positive mean. We test this observation on simulations with more

Table 2 Simulation parameters for Test 1: sensitivity analysis for chartists price biasing

Test 1		
Parameter	Value	Comment
μ_V	0 %	Fundamental value periodical average return
σ_V	1 %	1 % standard deviation
N	350	Number of simulated trading periods
Nr agents	500	(100−x)% Chartist proportion and x % Proportion of perfectly informed investors
Length of memory	3	Trading rules for chartists using the last 3 market returns
μ_C, σ_C	(1 %, 7 %)	Chartists use a mark-up of $R \in N(\mu_C, \sigma_C)$
Wealth	1,000, 10	Initial endowment with 10 stocks and 1,000 in cash

Table 3 Summary statistics for Test 1 with different proportions of investors

Percentage of informed investors (x %)	Percentage of chartist investors (1−x)%	Median bubble size	Average bubble size	Maximum bubble size
70	30	0	0.31	3.86
69	31	0	0.91	9.45
68	32	0	0.69	5.58
67	33	0.015	1.7558	14.069
66	34	0.04	4.42	40.13
65	35	1.19	5.49	38.115
64	36	1.91	9.306	56.182
63	37	3.59	8.6267	50.584
62	38	5.27	9.65	65.43
61	39	6.64	13.145	70.53
60	40	6.47	18.791	128.4

periods and observe a very weakly declining average wealth of chartist investors (depending on the mix of investors and evolution of the fundamental asset).

In the next tests we will look at the long term risk-adjusted returns of different strategies. We are interested in proving our second and third hypothesis, where SITH investors can make better profits than YODA or Chartist investors.

In our second test we challenge YODA value investors against Chartist investors. The parameters for this test are listed in Table 4. The proportions of investors are chosen rather arbitrary. The results remain qualitatively valid even if these proportions are substantially changed.

Even if YODA investors have biased information as individuals, in aggregate they receive unbiased fundamental information, therefore they do not create persistently biased price quotes. We run 100 simulations with Test 2 parameters and analyze the average final risk-adjusted return (computed using the Sharpe return ratio) for every type of investor.

We use as benchmark a buy&hold investment strategy which implies buying the asset at the beginning of the simulation and holding it until the end. The results in

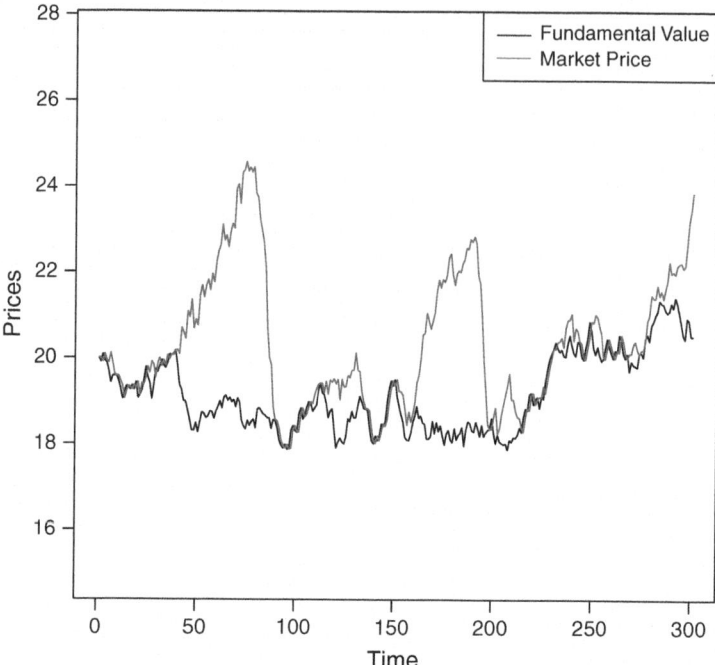

Fig. 1 Market price in a Test 1 simulation with 35 % chartists and 65 % perfectly informed traders

Table 4 Simulation parameters for Test 2: evolution of wealth

Test 2		
Parameter	Value	Comment
μ_V	0 %	Fundamental value periodical average return
σ_V	1 %	1 % standard deviation
N	500	Number of simulated trading periods
Nr agents	500	70 % Chartists informed investors 15 % perfectly informed YODA and 15 % biased informed YODA investors
a,b	$a \subset [0.5, 1.5]$ and $b \subset [-1, 1]$	Information bias parameters for biased YODA investors
Length of memory	3	Trading rules for chartists using the last 3 market returns
μ_C, σ_C	(2 %, 8 %)	Chartists use a mark-up of $R \in N(\mu_C, \sigma_C)$
Wealth	1,000, 10	Initial endowment with 10 stocks and 1,000 in cash

Table 5 Average portfolio returns per investment strategy from Test 2

Summary mean statistics for 100 simulations	Sharpe ratio	T-stat
Perfectly informed YODA investors	0.470085	6.219227
Biased YODA investors	**1.77264953**	**30.7699281**
Chartist investors	0.13588381	3.1067333
Benchmark buy&hold investors	0.14916993	4.89394981

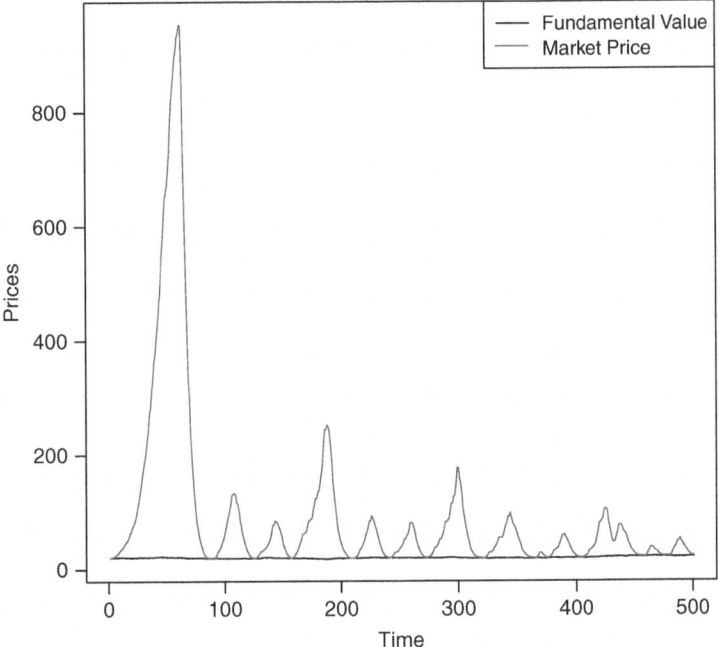

Fig. 2 Market price in a simulation with parameters from Table 4

Table 5 show that Chartist investors perform, on average, worse than the benchmark strategy. In Fig. 2 we observe bubbles that are created by the Chartist investors. Moreover, it is clear that the intensity of the bubbles diminishes as wealth is transferred to the informed YODA investors. As expected, Chartists lose money to informed YODA investors.[4] It is interesting to notice that among the YODA investors, it is the group with biased information that performs the best. Intuitively we think that better information implies better returns. This is true provided that markets reverse prices quickly towards what the fundamental value indicates. In our case, where the market is dominated by Chartist investors (70 %), prices revert slowly and infrequently. Biased YODA investors manage, as a group, to make more

---

[4]By setting the asset's fundamental growth to 0 % we ensure that investors can make money only at the expense of others, and not from fundamental growth.

Table 6 Simulation parameters for Test 3: evolution of wealth with all investor types

Test 3		
Parameter	Value	Comment
μ_V	0.01 %	Fundamental value periodical average return
σ_V	1 %	1 % standard deviation
N	500	Number of simulated trading periods
Nr agents	500	70 % Chartists informed investors 10 % perfectly informed YODA and 10 % biased informed YODA investors and 10 % biased SITH investors
a,b	$a \subset [0.5, 1.5]$ and $b \subset [-1, 1]$	Information bias parameters for biased YODA investors
Length of memory	3	Trading rules for chartists using the last 3 market returns
μ_C, σ_C	(2 %, 8 %)	Chartists use a mark-up of $R \in N(\mu_C, \sigma_C)$
Y	[30 %, 70 %]	SITH investors switch strategy, from Chartist to YODA, when the market mispricing is higher than Y
Wealth	1,000, 10	Initial endowment with 10 stocks and 1,000 in cash

Table 7 Results of Test 3

Summary statistics for 100 simulation	Sharpe ratio	T-stat
Perfectly informed YODA investors	0.97673179	10.231286
Biased YODA investors	1.97200353	27.2560436
SITH investors	**3.12170565**	**53.1889911**
Chartist investors	0.10076797	5.48089118
Buy&hold investors	0.12966009	7.5581478

transactions, and are able to close their positions faster than perfectly informed investors. This is due to the biased YODA investor's group heterogeneity which consists of a wide range of price expectations thus more chances of trading than the perfectly informed investors.

In our third test we introduce the SITH investors and a fundamental value with a small daily growth. The parameters of this test are listed in Table 6. SITH investors use a mix of technical and fundamental strategies. When the market price is too far from the fundamental value, and it passes over their respective Y threshold, SITH investors change their strategy and bet on the reversal of prices. The results of this market model are listed in Table 7. As in the previous test our results are robust to changes in any of the simulation parameters.

This third test shows that the best performing strategy belongs to the SITH investors. On the opposite end of the spectrum are the Chartist investors who perform even worse than the benchmark strategy. As in the second test, the biased informed YODA investors outperform the perfectly informed investors. When a price bubble develops, informed investors attempt to reverse prices and start selling in the following order: first perfectly informed investors (they immediately detect any mispricing), next are the biased informed YODA investors and the last ones to sell are the SITH investors (when they reach their threshold limit Y). During the bubbling period, SITH investors help push prices up and out of the other informed investors' portfolios. Because SITH investors ride and fuel bubbles, but also turn against the market when prices are too high, they manage to capture more wealth from Chartist investors (and from the fundamental growth) than do the informed YODA investors.

4 Conclusions

Within a simple model of a financial market we compare informed and trend-building strategies. We question the importance of perfect information by showing that better information does not necessarily provide better risk-adjusted returns. Using a comprehensive sensitivity test we show that, under certain conditions, a minimum of 35% technical traders is enough to persistently generate market price bubbles. Because the average wealth of chartist traders diminishes slowly the mispricing effect is enduring. SITH investors, which combine Chartist and fundamental strategies, have the best risk-adjusted returns. We state that informed investors have monetary incentives to contribute to price bubbles. When markets are close to efficiency, informed investors are better off using their information and betting on price reversals. Yet, when uninformed strategies dominate, it is more profitable for informed investors to mimic uninformed strategies (up to a certain threshold). From this tests we conjecture that financial markets are overestimating assets, most of the time. This overpricing is not due to risk premiums or high risk estimations but it is caused by the choice of a more profitable investing strategy, which often disregards the fundamental economic reality. In this light, we consider that financial markets are *seldom* efficient. These observations can be the basis of a behavioral theory of why financial markets are more and more volatile and information-to-noise levels are decreasing.

References

1. Allen F, Gale D (2000) Asset price bubbles and monetary policy. Center for Financial Institutions working papers, Wharton School Center for Financial Institutions, University of Pennsylvania, May, No. 01–26
2. Beaufils B, Brandouy O, Ma L, Mathieu P (2010) Simuler pour comprendre: un éclairage sur les dynamiques de marchés financiers à l'aide des systèmes multi agents. Systemes d'Information et management 14(4):51–70
3. Beja A, Goldman MB (1980) On the dynamic behavior of prices in disequilibrium. J Financ 35:235–248

4. Farmer JD, Joshi S (2002) The price dynamics of common trading strategies. J Econ Behav Organ 49:149–171
5. Grossman SJ, Stiglitz JE (1980) On the impossibility of informationally efficient markets. Am Econ Rev 70(3):393–408
6. Kahneman D, Tversky A (1979) Prospect theory: an analysis of decision under risk. Econometrica 47(2):263–291
7. LeBaron B (1999) Evolution and time horizons in an agent-based stock market. Macroecon Dyn 5:225–254
8. LeBaron B (2006) Agent-based computational finance. In: Tesfatsion L, Judd KL (eds) Handbook of computational economics, vol. 2. Elsevier, North-Holland, pp 1187–1232
9. LiCalzi M, Pellizzari P (2007) Simple market protocols for efficient risk sharing. J Econ Dyn Control 31(11):3568–3590
10. Muth JF (1961) Rational expectations and the theory of price movement. Econometrica 29(3):315–335
11. Toth B, Scalas E (2007) The value of information in financial markets: an agent-based simulation. Papers 0712.2687, arXiv.org

Emergence of Fair Offers in Ultimatum Game

Wanting Xiong, Han Fu, and Yougui Wang

Abstract The dynamics of how fair offers come about in ultimatum game is studied via the method of agent-based modeling. Both fairness motive and adaptive learning are considered to be important in the fair behavior of human players in concerning literature. Here special attention is paid to situations where adaptive learning proposers encounter responders with either pure money concern or fairness motivation. The simulation results show that the interplay of adaptive learning participants yields a perfect sub-game equilibrium, but fair offers will be provided by proposers as long as a small proportion of responders play "tough" against unfair offer.

1 Introduction

Simple and intriguing as it is, ultimatum game has been a widely applied analytical tool in bargaining behaviors between two parties who periodically bargaining in pairs over their shares of a common pie such as buyers and sellers or employers and employees. In a basic ultimatum game, there are two players and a pie. The proposer suggests how to divide the pie while the responder can either accept or reject it. If the deal is accepted, they share the pie accordingly. Otherwise, both players get nothing. Using a classic game theory method, we can approach equilibrium analysis with the following three postulations [1]:

P1: Each player's exclusive motivation is to gain the most money possible.
P2: Other bargainers' motivation is common knowledge.
P3: Each bargainer can identify his optimal action.

W. Xiong · H. Fu · Y. Wang (✉)
Department of Systems Science, School of Management, Beijing Normal University, Beijing 100875, People's Republic of China
e-mail: wanting.x12@gmail.com; fuhan3288@sina.com; ygwang@bnu.edu.cn

Since the proposer knows that the responder prefers more money to less (P1 and P2), the proposer should offer the responder the smallest monetary unit allowed, while allocating the balance to himself (P3). The responder should accept (P1 and P3). In this way we can identify many Nash equilibria.[1]

However, in a profusion of experiments, human players behave distinctly from theoretical predictions. The majority of proposers offer 40–50 % of the pie to their opponents, and about half of the responders reject offers below 30 % [2, 5, 6, 9, 10, 18]. This severe deviation of empirical results from the theoretical predictions is considered as a solid evidence of human's irrationality, and therefore is studied extensively in the economic literature. To explain human players' preference for fairness, two different types of theory prevail. One centers on fair motive, which postulates that human players are subject to no pure monetary incentives but a much richer motivation structure, including envy [11], inclination to reciprocate [13], inequality aversion [2, 3, 6, 15]. The other emphasizes adaptive learning and argues that fairness in those one-shot experiments is the reflection of a well-paid-off social habit formed in one's real life experience like bargaining [16]. In contrast to the former theory, the adaptive learning theory is often if not all studied in iterated game structure rather than one-shot games. Many models have been proposed such as optimal learning model [7, 8], reinforcement learning model [16], learning direction model [17] and evolutionary game model [12, 14]. Although in some lines the two hypotheses are considered competing, yet among others that underline the asymmetry in the behaviors of the proposer and responder, they are conceivably complementary. Roth et al. [16] pointed out that there is a significant disparity in the behaviors of proposers and responders which roots deeply in the nature of the game and gets amplified as the gainings of experience. With experiments designed to test the two hypotheses separately, Abbink et al. [1] found that the responder's rejection is more likely to come from one's fairness concerns while the proposer's generous offer is better explained as adaptive learning result from the game interactions.

To provide a simple and clear explanation which integrates both fairness motive and adaptive learning hypotheses, we employ an agent-based modeling approach to demonstrate how fair offers come about in ultimatum games. From the evolutionary perspective, we also attempt to discuss the formation process of a specific social norm in which some "tough" minority determines the fate of all. Throughout our work, we stress the displacement of hyperrational postulations of individuals by a more reasonable and less demanding description of agents as adaptive learning, profit maximizing and naive judging.[2] In this context, everybody reacts to the

---

[1] Any strategy for the proposers combined with the strategy to accept the offer and reject all lower ones is counted as a Nash equilibrium, and a unique subgame perfect equilibrium. Strictly speaking, in discrete cases where the strategy choice includes zero, it counted as a subgame perfect equilibrium with the proposer offering either zero or the smallest possible divisible piece to the responder.

[2] In other words, we replace P1 with P1* that all agents' exclusive motivation is to gain maximum amount of money except for some "tough" responders who follow a fairness motive and P3 with P3* that agents can identify his or her learning directions by comparing one's latest payoffs.

environment by forming a naive judgment on strategies upon one's own payoffs in the previous games while the real payoffs rest upon how other players respond.

With the notions given above, we propose here two multi-agent models to look into the dynamics of the repeated ultimatum game. In the basic model, both the proposers and responders are adaptive learners who are motivated by pure monetary incentives. It turns out that the system converges quickly to perfect sub-game equilibrium, irrespective of the initial conditions. Then we put forward a modified model including both distributive and fairness concerns by introducing a certain portion of "tough" responders who reject unfair offers regardless of their own monetary payoffs. The simulation arrives at two conclusions: the first is that the overall welfare of the responder group is increased owing to the rejections of the "tough" players to the profitable but unfair offers. The second is that even a surprisingly small population of the "tough" responders can induce the dominance of fair offers among the proposers. In order to better illustrate the emergence of fair offers, we also introduce "fairness leverage" to quantify the influence of the "tough" responders on the other group. Moreover, it is found through the comparison of the two models that the adaptive learning hypothesis alone cannot account for the fair outcome, yet together with fair motive settings, it can interpret the empirical evidences in both intermediate and long terms.

2 Basic Model

In this model, we concentrate on the dynamic process of the repeated ultimatum game, in which both the proposers and responders adjust their strategies incrementally by comparing payoffs they gained from their previous games. For simplicity, the step of the incremental adjustment, relatively small to the sum of money, is the same for all players. The group sizes of both parties are set to be N and the money stake for each game shot is M. The players' strategies are given by two parameters p and q: in each game, the proposer offers a proportion of the pie, p, to his partner while the responder rejects any offer lower than q. Then we have the offer set of the proposers, $P = \{p_i, i = 1, 2, \ldots, N\}$ and the demand set of the responders, $Q = \{q_j, j = 1, 2, \ldots, N\}$. A game period is defined as R rounds (usually R equals N) of randomly paired-up ultimatum games, in which each player meets R partners. We note that the players adopt a consistent strategy within one period and adjust it according to one's own profit calculation between any two adjacent periods. The payoff of the tth period, is $\Pi_i^p(t)$ for the proposer, and $\Pi_j^r(t)$ for the responder. For the first period, the offer and demand sets are initially subject to a uniform distribution over $[0, M]$. Then in the second period, the players will make a try by deviating a certain amount, δ (δ is much smaller than M), towards a random direction, from their initial strategy. From the third period, the proposers will compare their payoffs gained in the previous two periods and adjust their offer strategies by δ towards the same (opposite) direction of the last adjustment he made if the payoff increases (decreases). Mathematically, it is given as follows:

$$\text{if } \Pi_i^p(t) \geq \Pi_i^p(t-1) \text{ and } p_i(t) = p_i(t-1) + \delta, \text{ then } p_i(t+1) = p_i(t) + \delta;$$
$$\text{if } \Pi_i^p(t) \geq \Pi_i^p(t-1) \text{ and } p_i(t) = p_i(t-1) - \delta, \text{ then } p_i(t+1) = p_i(t) - \delta;$$
$$\text{if } \Pi_i^p(t) < \Pi_i^p(t-1) \text{ and } p_i(t) = p_i(t-1) + \delta, \text{ then } p_i(t+1) = p_i(t) - \delta;$$
$$\text{if } \Pi_i^p(t) < \Pi_i^p(t-1) \text{ and } p_i(t) = p_i(t-1) - \delta, \text{ then } p_i(t+1) = p_i(t) + \delta.$$
(1)

The responders' adjustments follow the same rule.

Along with all these assumptions and settings, we carry out the simulations and record the average number of successful deal, the average offer and demand, as well as the average payoffs for the proposers and responders respectively in each period. As shown in Fig. 1a, the amount of what the proposers offer and the responders demand both fall in the first 3,700 periods and eventually fluctuate around 7.5 ever since.[3] We regard the first 3,700 periods as the compromising stage where the slippery proposers gradually realize their first-move advantage and push the responders to lower their demands again and again. Figure 1b suggests that this adaptive learning mechanism does contribute to the overall welfare of the game participants by increasing the close rate in the game process. However, as indicated in Fig. 1c, there exists a significant inequality in the distribution of this growing pie, over 80 % of which is acquired by the proposers.

By changing the value of R, which determines how many partners one meets before he changes strategy, we take a further investigation on how the outcome of the game changes when agents base their strategic adjustment on partial information. As shown in Fig. 2a–c, the average demands of the responders in equilibrium state are practically identical while the average offers of the proposers at equilibrium increase from 5 ($R = N/10$), to 7 ($R = N/5$) and to 10 ($R = N$). The change implies that the quick adjustment with less information about the responders' reactions makes it harder for the proposers to push their offers down to the bottom as it amplifies the punishment from overly aspired responders who keep random walking at equilibrium. Also it can be drawn from Fig. 2 that both curves converge to equilibrium much slower as R decreases. To sum up, less information or uncertainty about others' reactions leads to higher offers from the proposers and a faster convergence to the agreement between the two parties.

---

[3]The reason why average demand of responders triumphs over average offer of proposers in the steady state is that some overly aspired responders keep random walking at a relatively high level as they turn down most offers.

Emergence of Fair Offers in Ultimatum Game

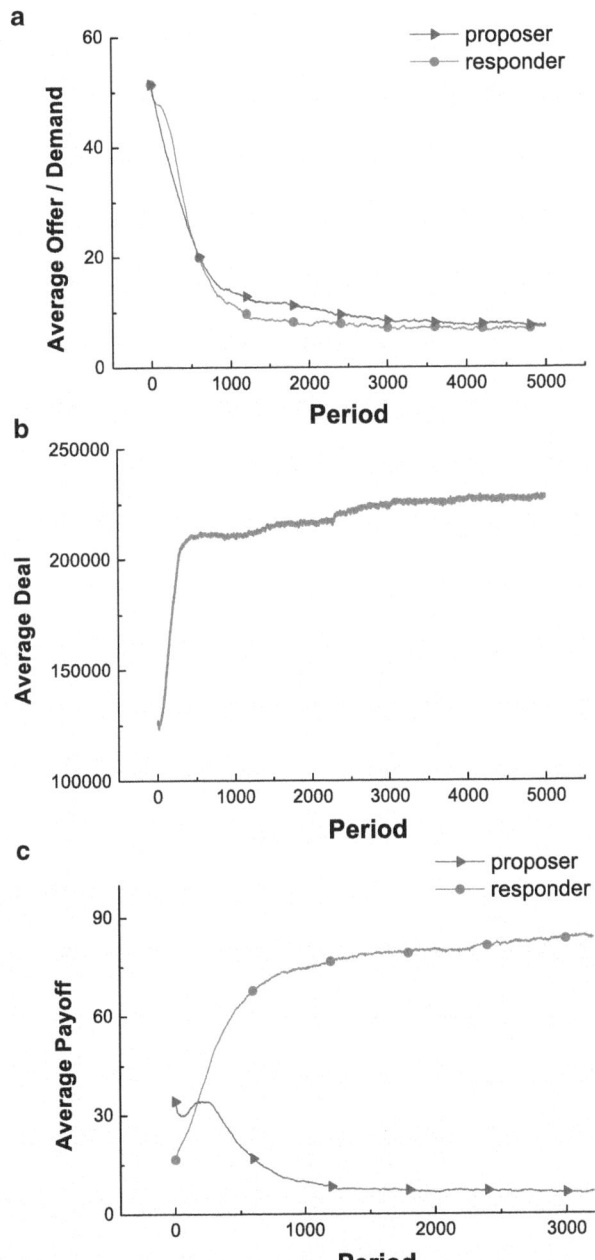

Fig. 1 The evolutions of (**a**) average offer and demand, (**b**) average deal, (**c**) average payoffs in repeated ultimatum game with $N = 500$, $R = 500$, $M = 100$, $\delta = 1$

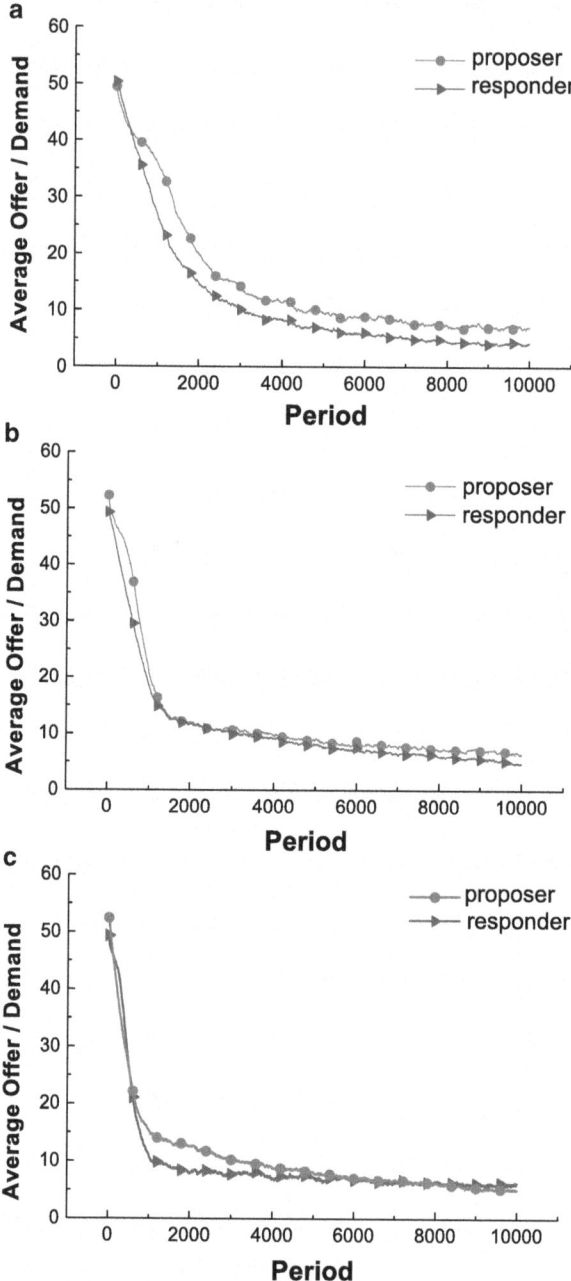

Fig. 2 The evolutions of average offer/demand with R = N/10 (**a**), N/5 (**b**), N (**c**)

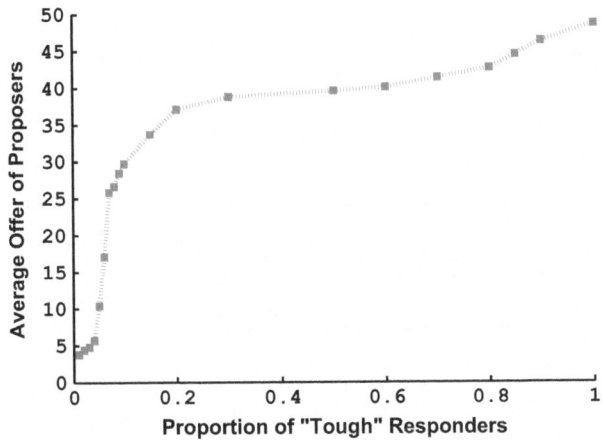

Fig. 3 The average offer of proposers at equilibrium versus the proportion of "tough" responders

3 Modified Model

To understand how fair offers emerge, we modify the basic model by introducing fair motive to the responders. Suppose that all responders are unhappy with their easily bullied strategy and decide to teach the proposers a lesson by rejecting all unfair offers (distribution plans other than 50/50) regardless of the short-term monetary loss. Then intuitively one would foresee that as a rational response the proposers will retreat to the fair outcome eventually. In reality, however, it is quite impossible that all responders are willing to sacrifice. So what will happen if only a fraction of responders are "tough" radicals?

To answer this question, we turn some responders into "tough" players who reject any offers lower than a threshold l while the others stick to the basic adaptive learning rule. For simplicity, we set the rejection threshold to be identical for all "tough" responders and repeat the experiment with different proportions of "tough" responders. The final results when the game arrives at the stable stage are shown in Figs. 3–5.

As shown in Fig. 3, the average offer starts with a sudden rise and then go much slower as the proportion of "tough" responders increases. With only 10% responders playing "tough", the average offer of the proposers amounts to almost 30. This conclusion is confirmed by Brenner's experiment [4] that when human proposers repeatedly encounter artificial responders with the assigned rejection strategy, a higher acceptance rate or lower demand leads to lower offers.

Meanwhile, Fig. 4 displays the payoffs at equilibrium of both proposers and responders when facing different proportions of "tough" responders. As the proportion of "tough" responders increases, the average payoff of responders at equilibrium goes up abruptly and then slides down slightly while the payoff of the

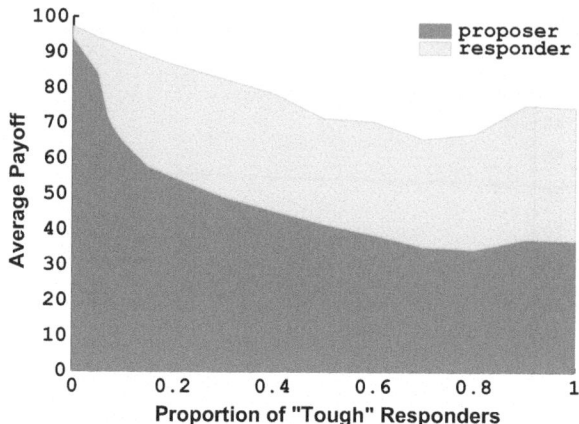

Fig. 4 The average payoffs of proposers, responders and the whole population at equilibrium versus the proportion of "tough" responders

proposers falls and bounces up accordingly. Although playing "tough" suffers a monetary loss for individuals in a short period of time, the responder group does benefit from the induced fair offers from the other group. However, as shown in Fig. 4, the average payoff of all game players overall declines as the responders become tougher, which suggests that the victory of fairness does not come about without any sacrifice: the overall welfare, or the pie shared by the two parties becomes actually smaller as "toughness" gradually prevails.

Similar conclusions can be drawn from Fig. 5, in which we compare the evolution processes of payoffs with different proportions of "tough" players. The upward sloping curve of proposers reflects the rising propensity of the proposers' bargaining power through trial and error and the downward sloping curve of responders represents the group's retreat. Nevertheless, thanks to those "tough" players, the responder group manages to end up with a relatively fair payoff at equilibrium. Moreover, the different speed of convergence to equilibrium in the situations with different "toughness" degree implies that the "tough" strategy of responders is a more confirmative threat, which can be perceived by the proposers more quickly, than the adaptive learning strategy.

Figure 6 provides a more clear evidence of the leverage effect of toughness on the proposer's offer by counting the percentage of "fixed" proposers whose offer sticks around the "tough" responders' threshold. When the proportion of "tough" responders are too low (below 4 %), no proposers are locked up. However, fair offers emerge when "tough" responders add up to 5 % of the group and the fair trend grows rapidly to become the mainstream even when the proportion of "tough" responders is still relatively small (5–30 %). In conclusion, with only a small fraction (about 30 %) of the responders playing "tough", a surprisingly large percentage (about 63 %) of the proposers provide a fair offer.

Emergence of Fair Offers in Ultimatum Game

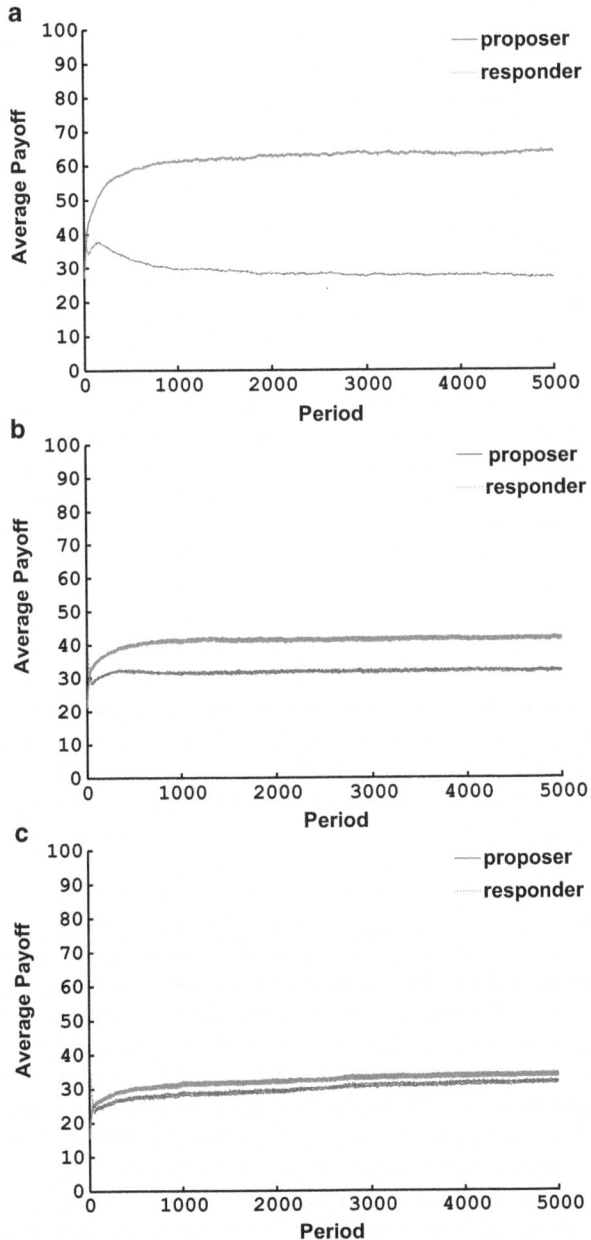

Fig. 5 Average payoffs of proposers and responders over time with proportion of "tough" responders equal to 10% (**a**), 50% (**b**), 80% (**c**)

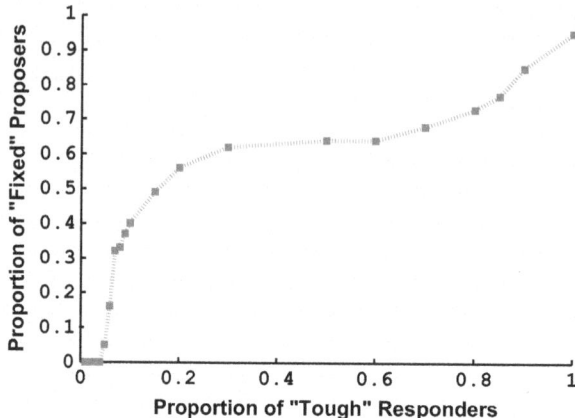

Fig. 6 The proportion of the "fixed" proposers in the steady state versus the proportion of the "tough" responders

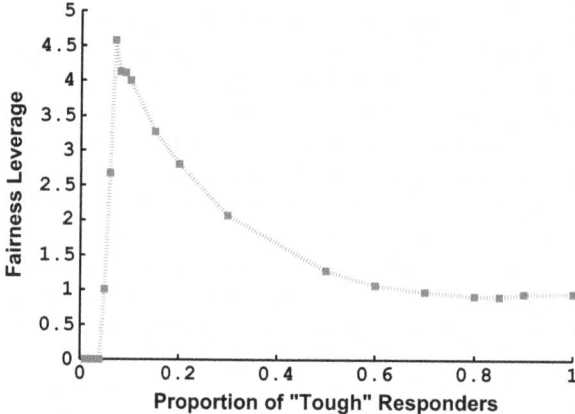

Fig. 7 The fairness leverage versus the proportion of the "tough" responders

To quantify the degree of "fixing" effect of the "tough" responders on the proposers, we introduce the concept of "fairness leverage", or the marginal "fixing" ability, which is calculated as the ratio of the number of "fixed" proposers to that of "tough" responders in equilibrium state. If the value of the fairness leverage is greater than one, it means that one "tough" responder is able to make more than one proposers to provide fair offers around his rejection threshold. As illustrated in Fig. 7, the curve of fairness leverage shows a horizontal start with its value remaining 0 until the percentage of "tough" responders reaches 5 %. This indicates that when the "tough" population is too small, no proposers care about their self-harming rejections as a sufficiently large number of responders (over 95 %) respond to their unfair offers in a bear and forebear manner. The curve then rockets up to its peak at 4.57 when the "tough" responders take up 7 % of the population and plunge down to stable around 0.9 afterwards. We point out that the majority values of fairness leverage (when the proportion of "tough" responders is within the range of 4–7 %) are greater than one, which implies that during most of the time one "tough" responder's rejection behavior entraps more than one proposers into the fair springe. This leverage effect is firstly initiated by the "tough" responders and then amplified through the following cycle: the rejections of the "tough" responders make some proposers offer more, and then the growing generosity of these proposers induces higher demands among the adaptive learning responders, whose rejections in turn attribute to more fair offers. Along with the increasing "toughness", the cycle repeats and the system eventually converges to the equilibrium.

4 Conclusion

In this paper, we studied the dynamics of how fair offers come about when players are subject to both cognitive and motivational irrationality. Using the basic model where the players are adaptive learners with only monetary concerns, we find that when encountering adaptive learning responders will be forced to lower demands repeatedly due to their myopic directional learning strategy, and the proposers, discovering the feebleness of the other group through interactions, will provide offers that converge to sub-game perfect equilibrium. In the modified model, we exhibited the correlation between proposer's fair offer and responders' appeal for fairness by turning a proportion of responders into "tough" players who reject unfair offers regardless of their own monetary loss. In accordance with the result of Brenner's experiment [4], we demonstrated that more intensified and tough demand for fairness would lead to higher offers from the proposers since the willingness for fairness are perceived by responders as a lower acceptance punishment. Moreover, we found that fair offers will emerge and eventually dominate among the proposers with a surprisingly small proportion of responders playing "tough". Other detailed analyses are also made concerning the relation between payoffs at equilibrium and the proportion of "tough" responders as well as the evolutions of payoffs over time, which generate similar results.

In spite of the fact that we provide no experiments but pure theoretical model in this study, evidences and applications of our conclusions can and shall be found in a variety of real-life examples. Suppose there is a job market where the employers are endowed with the first mover advantage in proposing salaries (offer) and the employees respond by deciding whether to take the job or not, the size of the pie is the monetary expression of the value created by the employee's labor. If the employees take any offer their employers offer, the profit maximizing firms will constantly lower down the salary until it reaches the possible minimum. However, the reality contrasts this miserable scene as some radical employees will go on strikes to protest against the firm's unfairness. Of course, "tough" strikers will not get paid by rejecting unfair offers. Whether the employees can achieve a fair victory or not usually depends on how long the strike lasts and how many people are involved. If the strikers stick to the "tough" strategy forever like what we presume in our model, it is intuitive that it would take a relatively small proportion of employees rather than the whole population to make the employers raise their offer. Similar story can be found between buyers and sellers where buyers boycott against insanely high prices.

Acknowledgements We are grateful to Dr. Jianzhong Zhang for his suggestions and language improvements. This research was supported by National Natural Science Foundation of China under Grant of No. 61174165 and Program for New Century Excellent Talents in University (NCET-10-0245). This work was also the result of Interdisciplinary Salon of Beijing Normal University.

References

1. Abbink K, Bolton GE, Sadrieh A, Tang FF (2001) Adaptive learning versus punishment in ultimatum bargaining. Games Econ Behav 37:1–25
2. Bolton GE (1991) A comparative model of bargaining: theory and evidence. Am Econ Rev 81:1096–1136
3. Bolton GE, Ockenfels A (2000) ERC: a theory of equity, reciprocity, and competition. Am Econ Rev 90:166–193
4. Brenner T, Vriend NJ (2006) On the behavior of proposers in ultimatum games. J Econ Behav Organ 61:617–631
5. Fehr E, Gachter S (1999) Cooperation and punishment in public goods experiments. Institute for Empirical Research in Economics working paper no. 10; CESifo working paper series no. 183
6. Fehr E, Schmidt KM (1999) A theory of fairness, competition, and cooperation. Q J Econ 114(3):817–868
7. Gittins JC (1979) Bandit processes and dynamic allocation indices. J R Stat Soc Ser B 41:148–177
8. Gittins JC (1989) Multi-armed bandit allocation indices. Wiley, Hoboken
9. Guth W, Tietz R (1990) Ultimatum bargaining behavior: a survey and comparison of experimental results. J Econ Psychol 11:417–449
10. Guth W, Schmittberger R, Schwarze B (1982) An experimental analysis of ultimatum bargaining. J Econ Behav Organ 3:367–388
11. Kirchsteiger G (1994) The role of envy in ultimatum games. J Econ Behav Organ 25:373–389
12. Nowak MA, Page KM, Sigmund K (2000) Fairness versus reason in the ultimatum game. Science 289(5485):1773–1775
13. Rabin M (1993) Incorporating fairness into game theory and economics. Am Econ Rev 83:1281–1302
14. Rand DG, Tarnita CE, Ohtsuki H et al (2013) Evolution of fairness in the one-shot anonymous ultimatum game. Proc Natl Acad Sci 110(7):2581–2586
15. Rotemberg JJ (2008) Minimally acceptable altruism and the ultimatum game. J Econ Behav Organ 66:457–476
16. Roth AE, Erev I (1995) Learning in extensive-form games: experimental data and simple dynamic models in the intermediate term. Games Econ Behav 8:164–212
17. Selten R, Stoecker R (1986) End behavior in sequences of finite prisoner's dilemma supergames: a learning theory approach. J Econ Behav Organ 7:47–70
18. Thaler RH (1988) Anomalies: the ultimatum game. J Econ Perspect 2:195–206

Part IV
Financial Markets

An Agent-Based Investigation of the Probability of Informed Trading

Olivier Brandouy and Philippe Mathieu

Abstract We study the Volume Synchronized Probability of Informed Trading (VPIN) proposed by Easley D, López de Prado M, O'Hara (Rev Financ Stud 25:1457–1493, 2010) as a consistent measure of the "order flow toxicity". The VPIN is a proxy for the probability that informed traders adversely select uninformed ones, notably Market Makers. We use a price-driven, asynchronous, agent-based artificial market where populations of agents evolve according to the general logic and within a similar framework as proposed by Easley D, Kiefer D, O'Hara M, Paperman J (J Financ 51(4):1405–1436, 1996). Among others, we document situations in which the VPIN is at high levels even if no informed trading is at play. This *ambiguity in the consistency* of the VPIN suggests that this measure may mislead competitive market makers in their decisions about the spread.

1 Introduction

Numerous recent market events, such as the Flash crash of May 6th 2010, have reactivated the debate around the possible deleterious effects of *informed trading* on market dynamics. Informed trading occurs when an investor uses a private information to trade financial assets at interesting prices with regard to the normal price, should this information be public. When such informed trading occurs, the winner is the informed trader, and the loser is the market maker who has the obligation of permanently offering bid and ask quotes for the securities listed

O. Brandouy (✉)
Sorbonne Graduate Business School, 75005 Paris, France
e-mail: olivier.brandouy@univ-paris1.fr

P. Mathieu
LIFL, UMR CNRS-USTL 8022, Université Lille 1, 59100 Lille, France
e-mail: philippe.mathieu@lifl.fr

in the market. Identifying such dangerous situations is thus of major interest, notably for market makers. For Easley et al. [4] the "flow toxicity" coming from the activity of informed traders can be monitored in the market with a metric called "Probability of INformed trading" (a.k.a. PIN).

In a more recent paper, Easley et al. [3] have developed a "Volume Synchronized Probability of Informed Trading" indicator (VPIN), which merely captures in real time the potential flow toxicity in high frequency markets. The VPIN may be regarded as a practical implementation of the PIN. As the order flow toxicity increases, the less informed market makers have to withdraw so as to avoid further losses. In doing so, they may be drawing even more liquidity out of the market and increase the overall level of flow toxicity in the traded volume.

This vicious cycle mechanism eventually achieves in forcing all the market making activity out of the market and consequently, may induce liquidity crashes.

However, since it is practically impossible to access all the relevant information to calculate both the PIN and the VPIN, these latter may be subject to rightful interrogations.

In this paper, we analyse the validity of the VPIN in an artificial agent-based stock market. To the best of our knowledge, this is a first attempt to investigate the properties of the VPIN using this artificial-intelligence methodology.

This paper is organised as follows. Section 2 exposes how the VPIN can help in detecting toxic orders. The agent-based platform, the artificial agents behaviors and the experimental plan as well are detailed in Sect. 3. Results and their discussion are presented in Sect. 4.

2 The Easley and al. Models[1]

The PIN was first defined by Easley et al. [1]. The fundamental idea in this paper is to model the price intervals in which a risk neutral market maker accepts to provide liquidity. The basic structure of a trading day, in which the model is developed is proposed in Fig. 1.

In this model, time is continuous and traders (both informed or uninformed) may enter the market at any moment. This flow of investors is modelled by two independent Poisson processes with parameter μ for the informed and ε for the uninformed. Traders find in the positions of risk-neutral, competitive market makers (here-after MM) a counterpart. The latter have to propose Bid and Ask quotes continuously, and investors make their decisions using these quotes and their information regarding the value of the (single) traded asset. Since MM are risk-neutral and competitive, these quotes reflect their expectation about the value of the traded asset.

Each day a new information may be disclosed with a probability α (the game is developed over several days and learning is possible). This signal is either good (with a probability $(1-\delta)$) or bad (probability δ). At the end of the day, the full value

---

[1] Note that the models presented in this section are not developped by the authors but are explicitely borrowed from a set of papers by Easley et al. [1–3 and 4].

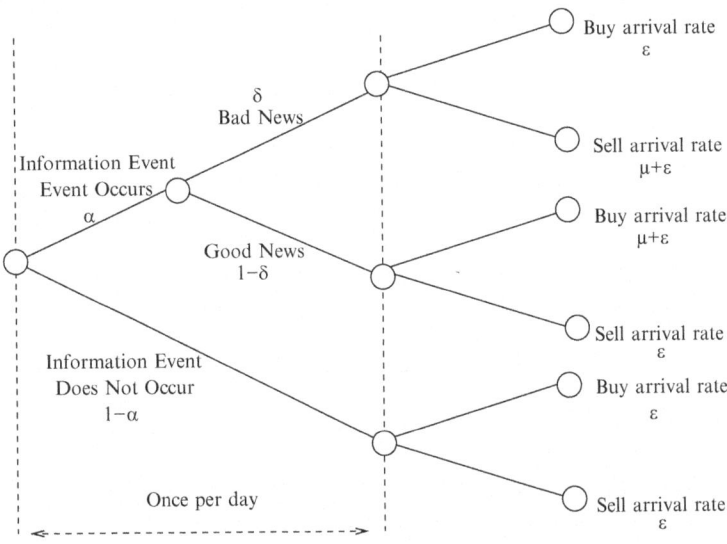

Fig. 1 Trading process diagram, see Easley and al. [1] (fig.1, p.1409)

of the asset is realized (S_G if a good news was drawn, S_B in case of a bad news; S is the value of the asset if no information is disclosed). By assumption $S_B < S < S_G$.

If uninformed traders are present whenever information is drawn or not, informed traders only send orders in case they can benefit from their informational advantage.

MM do know the structure of the tree and the probabilities in Fig. 1. Nevertheless, they do not know if a good or a bad information (or no information) has been drawn. Thus, according to these authors, the expected value of the security at the beginning of the day is:

$$E[S_t] = (1 - \alpha_t)S + \alpha_t [\delta_t S_B + (1 - \delta_t)S_G] \qquad (1)$$

Easley et al. [1] assume that MM are Bayesian updaters learning from the rate of arrival of orders if a Good or a Bad information governs the order flow. From these elements, they show that to avoid losses (i.e. to be counterpart of an informed trader), the MM would propose the following quotes:

$$E[B_t] = E[S_t] - \frac{\mu \alpha_t \delta_t}{\epsilon + \mu \alpha_t \delta_t}(E[S_t] - S_B) \qquad (2)$$

$$E[A_t] = E[S_t] + \frac{\mu \alpha_t (1 - \delta_t)}{\epsilon + \mu \alpha_t (1 - \delta_t)}(S_G - E[S_t]) \qquad (3)$$

Easley and al. [1] show that, combining Eqs. 2 and 3 delivers an expected Bid-Ask-Spread:

$$E[A_t - B_t] = \frac{\mu \alpha_t (1 - \delta_t)}{\epsilon + \mu \alpha_t (1 - \delta_t)}(S_G - E[S_t]) + \frac{\mu \alpha_t \delta_t}{\epsilon + \mu \alpha_t \delta_t}(E[S_t] - S_B) \qquad (4)$$

If one accepts that δ_t is equal to 0.5, it turns that Eq. 4 can be simplified

$$E[A_t - B_t] = \frac{\alpha_t \mu}{\alpha_t \mu + 2\epsilon} (S_G - S_B) \qquad (5)$$

In Eq. 5, $\frac{\alpha_t \mu}{\alpha_t \mu + 2\epsilon}$ is the Probability of Informed Trading (PIN_t).

Thus, the arrival of uninformed and informed traders affect prices (and quotes, subsequently). If $\mu = 0$, trades do not reveal any information and quotes simply reflect the prior expected value of the asset. If $\epsilon = 0$, then $E[B_t] = S_B$ and $E[A_t] = S_A$ which will avoid any trading. If a mixed population of traders exists (which is probably the more likely), the spread reveals how MM protect themselves from informed trading.

However, when setting up their spread intervals and trying to avoid the toxicity flows, the market makers do not know the parameters α, δ, μ and ε. Thus, according to Easley and al. [3] they have two possibilities to estimate this metric:

1. Estimating the parameter vector $\theta = (\alpha, \delta, \mu, \varepsilon)$ which is a far more complex task than estimating rates from independent Poisson processes. This can be done using maximum likelihood estimation.
2. In a high frequency world, the problem is even more complex since the quotes proposed by the MM must be adapted in continuous time and with a very low latency. However Easley et al. [4] have shown that the maximum likelihood estimation can be replaced by a fairly easier-to-compute metric denominated VPIN. In their paper, they show that the VPIN is a good estimate of the PIN (see also Easley et al. [2]). Its computation is made through a simple aggregation of the signed exchanged volumes in the market. To cite Easley and al. [3],

> VPIN is updated in volume-time, making it applicable to the high frequency world, and it does not require the intermediate estimation of non-observable parameters or the application of numerical methods.

The logic of the computation of the VPIN is as follows:

- Let V_θ^S and V_θ^B be the volumes traded against the Ask side and respectively the Bid side of the order book.
- The probability that the flow contains toxic orders coming from the informed traders can be estimated by the ratio of (i) the orders emitted by informed traders when private information arrives over (ii) all the volume generated by the overall activity in the market. (i) and (ii) can be estimated from the Eqs. 6 and 7, (see Easley and al. [3]).

$$E[|V_\theta^S - V_\theta^B|] \simeq \alpha \mu \qquad (6)$$

$$E[|V_\theta^S + V_\theta^B|] = \alpha \mu + 2\varepsilon \qquad (7)$$

Therefore, for these authors a good PIN estimator is given by the Eq. 8.

$$VPIN = \frac{E[|V_\theta^S - V_\theta^B|]}{E[|V_\theta^S + V_\theta^B|]} \qquad (8)$$

Notice that, unlike most of the empirical works scrutinizing the PIN and the VPIN (notably Easley and al. [3]), since our empirical methodology rely on an artificial market where all Bid and Ask orders may be archived and processed, we do not need to use a Lee-Ready algorithm Lee [5] to classify trades depending upon the fact they are triggered by a Bid or a Ask order.

We now move to the empirical design of the paper.

3 Agent Based Simulations and Empirical Tests

We design experiments to understand whether the VPIN metric is accurate in spotting the presence of informed traders. For doing so, we implement, in an artificial stock market described later on, the following protocol:

- We mix two different types of virtual agents: zero-intelligence traders (ZIT) who stand for the liquidity providers, and Informed Traders (IT) which try to maximize their profits.
- We start with a non-toxic environment encompassing only liquidity providers and we substitute progressively part of this population by informed traders. The overall population remains constant while the proportion of informed traders increases by a fixed percentage of the population. Information is disclosed for all the experiments at round 50 in the continuous trading period.

We collect the time series of prices, volumes, orders, the make-up of our trading panel as a mix of informed and uniformed traders, and the VPIN values over the experiments. Using these data, we observe the increase of informed trading through the VPIN metric and its reactivity.

Since we rely on an agent-based simulation method itself using non deterministic procedures, we run 100 replications of the experimental treatments so to obtain a good estimate of the average outcomes.

All the simulations are based on a simulated trading day. Prices, volumes and all necessary informations are collected on a tick-by-tick frequency basis. The simulation platform grants the possibility to follow, archive and eventually analyse the entire order book on the fly.

3.1 The Platform

In this research, we use an artificial market simulator called "ArTificial Open Market" (here-after ATOM, see Mathieu [7]). In this simulator, price formation directly emerges from the interaction of agents at the micro-level. The market microstructure in ATOM is duplicated from the Paris NYSE-Euronext stock exchange with the following features[2]:

---

[2]For technical details describing the simulation platform, see Mathieu [7].

1. It can emulate several order-books in parallel. Each order-book runs a continuous, double auction mechanism.
2. All types of NYSE-Euronext orders are allowed: limit, market, cancel or update orders, as well as sophisticated combinations such as stop-limit orders or limit orders with "*iceberg*" execution.[3]
3. The philosophy ruling the platform is such that an agent is an abstract entity that can be instantiated both by an autonomous process (developing its own strategy), or through an interface allowing humans to interact with the system. This latter possibility is not used in this research.
4. Concerning virtual agents, their possible behaviour is flexible and can be designed to fit the researchers' requirements. At each time step, the scheduler system offers agents the possibly to decide if they will send or not a new order for each of their stocks to the corresponding order book.

Particular attention is paid to the accuracy of calculations made by the platform so as to ensure the trustworthiness of prices it delivers. Notably, no real types are used or arithmetic divisions are allowed in the code. A logging system systematically historicizes all agents' decisions, as well as their impacts on the market. This is an essential feature to exploit the "*replay-engine*" ability of the simulator: if a population of agents simply re-processes real orders submitted to the market on a given day, the resulting price series is exactly the same as the one produced by the market.

3.2 Agents

Three kind of artificial agents are implemented in ATOM: Zero Intelligence agents (here-after ZIT), Informed Traders (ITR) and one Market Maker (MM).

ZIT   mostly use random number generators to determine prices and volumes in their orders. These agents do not rely on artificial intelligence methods such as classifiers or learning mechanisms to adapt their behaviour and/or to evolve. For the sake of possible replication of the results presented in this research, the pseudo-code describing these agents is available in Appendix. These agents are directly inspired by the work of Maslov [6]. The common characteristics for all of our ZITs are as follows:

- They only send limit orders.
- Each agent can submit both orders, Buy and Sell.
- Buy and Sell orders arise with equal probability ($p = 0.5$).
- Quantities are drawn in the range [10, 100].

---

[3]See NYSE-Euronext rule-book, at http://www.euronext.com. In this paper we only use "Limit", "Market" and "Cancel" orders.

ITR   receive during the experiment an information consisting in the "true" value of the traded asset. This value may either be 149.5 (when the information is "good") or 140.5 (when it is "bad"). Prior to receiving this information, ITR do not send orders to the market. As soon as they receive this information, they try to use it efficiently. Their basic behaviour can be summarized as follows (the pseudo-code describing these agents is available in Appendix):

- Like ZITs, they only send limit orders.
- If the information is "good", consider the best Ask price. If this price is lower than the "true" value (which should be the case), send a Bid order with a reservation price randomly drawn in the range [Best Ask, True value]. Conversely, if the information is "bad" send a Ask order with a reservation price randomly drawn in the range [True value, Best Bid].

MM   is an agent geared at offering a reasonable level of liquidity in the market, whatever the actual level of activity coming from the flow of orders emitted by the investors. An overview presenting some important features of electronic market making can be found in Nevmyvaka [8].

- The MM strategy simply consists in obtaining the best position in the Bid Ask queue.
- The MM continuously proposes, at the best limits, an arbitrary quantity equal to $\zeta = 1.1$ times the maximum volume available at the corresponding second best limit. If no such second best limit exists, this quantity is set at a level of 100.

Thus:

- Time is discretized into steps and the simulated trading is divided into 600 rounds in the platform. Ten rounds are dedicated to the pre-opening session (delivering the *first price of the day* after a batch auction fixing), 10 being used for a similar closing session ending-up with the daily closing-price. Thus, the continuous trading session lasts only 580 rounds.
- The total number of agents does not vary over time and is equal to 100. However, the number of agents within each category (informed/uninformed) is a control variable of the experimental treatments and is modified accordingly.
- There is, in addition to this population, only one Market Maker who constantly offers bid and ask quotations. The basic microstructure of the market is thus "*price driven*" and very similar to many real-world markets.
- All traders receive, at the beginning of each experiment, the whole set of the prices that might occur at the end of the trading session depending upon the arising of a new state of Nature.
- Information is always displayed at round 50 of the continuous trading period. Note again that no more than one information can arrive during an experiment. $\delta = 0.5$ which means that there are equal chances to be in a good or in a bad state of Nature. If no information occurs, then the asset value remains equal to $S|(\Omega = \emptyset)$ independently of the current traded price P_t.

- These different states impact the "true" value of the traded asset according to the following rule:

$$S|(\Omega \neq \emptyset) \begin{cases} \bar{S} & \text{In case of a Good event} \\ \underline{S} & \text{In case of a Bad event} \end{cases} \quad (9)$$

Ω in Eq. 9 is the information set.

Since liquidity providers do not know if an information has arrived, the range they use to draw their order prices is $[(1-\gamma)\underline{S},(1+\gamma)\bar{S}]$, γ being an idiosyncratic variable randomly chosen, for each of these agents, in $U(0, 1)$. They may be seen as Zero Intelligence Traders (ZIT).

Informed traders (ITR) do know in which state Nature will end up as soon as a new information comes in. Thus, they have an unfair advantage over the others. At each time step t in a trading session, they compare the current spread (B_t, A_t) with the "true" value of the asset at the end of the trading period $S|(\Omega \neq \emptyset)$. If the true value is within the spread they cannot exploit this private information to make profits. However, if this is not the case profits can be captured using this discrepancy.

Recall we want to verify if the VPIN is accurate in spotting the presence of informed traders. We thus focus on the level of the VPIN and the quantity of informed traders in the market to assess if this statistics fairly signals these situations or not.

4 Results and Discussion

As said previously, we increase the potential market impact of informed traders in substituting at each step, and from a homogeneous non-toxic population of liquidity providers (or ZIT agents), one ZIT with one informed trader (ITR) and observe how the VPIN evolves along this treatment. Note again that informed traders remain out of the market (in the sense they do not send any order) prior the disclosure of information, which systematically takes place at round 50. In Fig. 2 we present the evolution of the mean (Fig. 2a) and standard deviation (Fig. 2b) of the observed VPIN in each run of the experiment.

The mean VPIN clearly tends to increase with the number of informed traders which suggests it spots an anomaly in the market. However, it also reaches promptly a relatively high level of volatility. This suggests that even in situations where informed traders populate the market they may not be spotted accurately by the VPIN. The overall picture of the idiosyncratic evolution of the VPIN per run is proposed in Fig. 3.

In Fig. 3, the X-axis indexes the different runs and the Y-axis indexes the observations of the VPIN in each run. Patches of different colors indicate the level of the VPIN (see scale on the right side of the figure). Since ITR only start

An Agent-Based Investigation of the Probability of Informed Trading

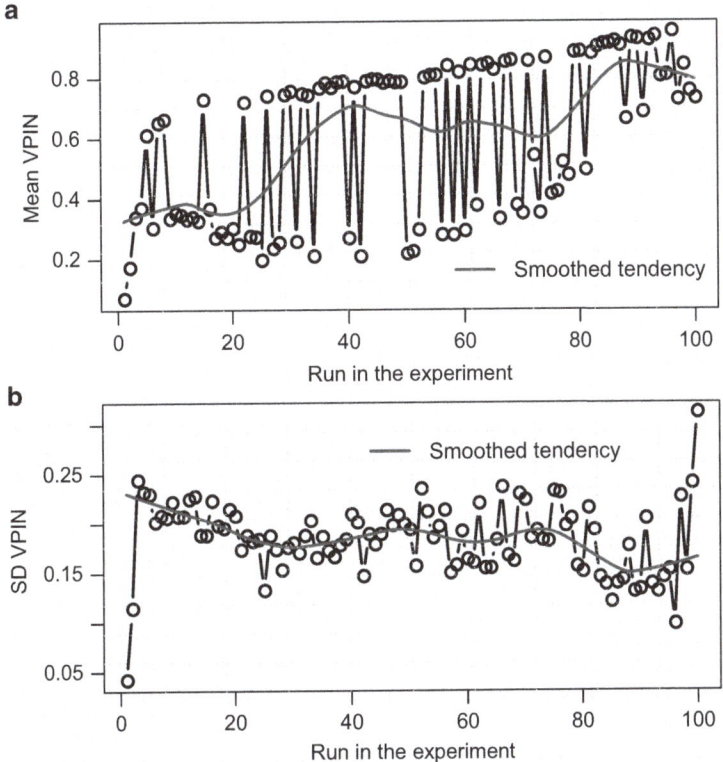

Fig. 2 Effect on the VPIN of an increasing population of ITR. (**a**) Mean of the VPIN. (**b**) Standard deviation of the VPIN

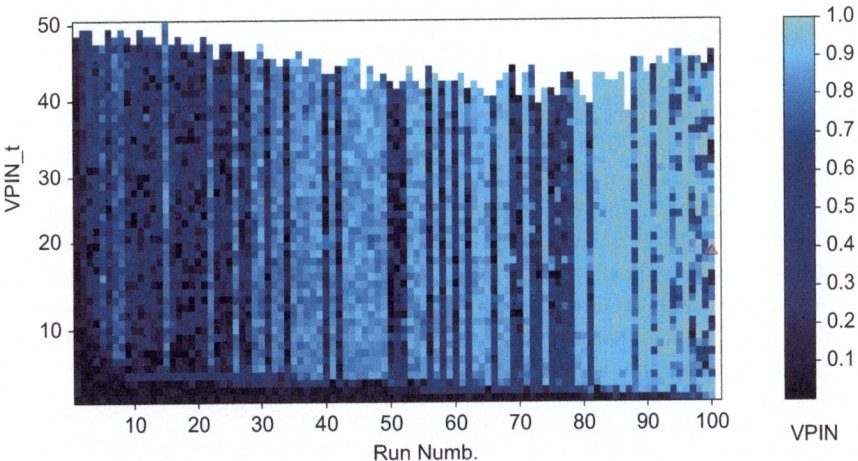

Fig. 3 Dynamics of VPIN, 100 runs

trading after round 50 where the information is disclosed one should consider that index corresponding to the moment where their impact may be visible in the figure decreases with their number in the runs (and consequently, in the number of analyzed runs).

The mean and the standard deviation of the VPIN is calculated using the 100 series of VPINs calculated at each run (there are between 46 and 50 such VPINS in each series).

The first run is characterized by a homogeneous ZIT population. This population becomes more and more heterogeneous when moving to the right side of the figure until run 50 where the highest heterogeneity is realized (50 ZIT and 50 ITR). This heterogeneity decreases when moving from run 50 to 100 where the population is again homogeneous, but only made of ITR except one liquidity provider. The first observation that can be made on Fig. 3 is that there is not an homogeneous impact on the VPIN when the number of ITR increases. Two cases may be identified:

1. At both the right and the left hand side of Fig. 3, the level of the VPIN seems either very low (dark patches indicating a VPIN between 0 and 0.5) or very high (light patches indicating a VPIN between 0.5 and 1). However, it is not impossible at all that among these observations, some VPIN tend to induce a type I error (concluding an absence of ITR although they do trade in the market).
2. In the middle of the figure appears a very heterogeneous picture where it can hardly be concluded that VPIN actually spots without ambiguity the presence of ITR. For one combinations of ZIT and ITR to another, the overall conclusion seems opposite. Consider for example runs around 45 (56 ZIT against 44 ITR) where high levels of VPIN are observed and runs around 75 (26 ZIT against 74 ITR) where relatively low levels can be pointed out.

5 Conclusion

We investigate the VPIN proposed by Easley et al. [3] as a measure of the "order flow toxicity". The VPIN is a proxy for the probability that informed traders adversely select uninformed traders, and notably Market Makers. We use a price-driven, asynchronous, agent-based artificial market where populations of agents evolve according to the general logic and within a similar framework as proposed in the seminal paper of Easley et al. [1]. We observe situations in which the VPIN is at high levels even if no informed trading occur. This *ambiguity in the consistency* of the VPIN suggests that such high levels may mislead the decision of a Market Maker only relying on this metric.

Appendix: ZIT and ITR Behaviour

Data: $\gamma, P_{min}, P_{max}; V_{min}, V_{max}$
Result: Order
/* initialisation   */
$\Lambda \sim U(0,1)$
/* equal possibilities to buy or sell   */
if $\Lambda > 0.5$ **then**
| $Direction = $ "Ask"
else
| $Direction = $ "Bid"
end
/* price and quantity definition   */
$P \sim U((1-\gamma)P_{min}, (1+\gamma)P_{max})$
$Q \sim U(V_{min}, V_{max})$

return $(Direction, P, Q)$

Algorithm 1: A generic zero intelligence trader (*ZIT*)

Data: $\alpha, S_B, S_G; V_{min}, V_{max}$
Result: Order
/* initialisation   */
$\alpha \sim U(0,1)$
/* Information is disclosed or not   */
if $\alpha > 0.5$ **then**
| $Direction = $ "Bid"
| /* price and quantity definition   */
| $P \sim U(P_{min}, S_G)$
| $Q \sim U(V_{min}, V_{max})$
else
| $Direction = $ "Ask"
| /* price and quantity definition   */
| $P \sim U(S_B, P_{max})$
| $Q \sim U(V_{min}, V_{max})$
end
return $(Direction, P, Q)$

Algorithm 2: A generic informed trader (*ITR*)

References

1. Easley D, Kiefer D, O'Hara M, Paperman J (1996) Liquidity, information and infrequently traded stocks. J Financ 51(4):1405–1436
2. Easley D, Engle F, O'Hara M, Wu L (2008) Time-varying arrival rates of informed and uninformed trades. J Financ Econom 6(2):171–207

3. Easley D, López de Prado M, O'Hara (2010) Flow toxicity and liquidity in a high-frequency world. Rev Financ Stud 25:1457–1493
4. Easley D, López de Prado M, O'Hara (2010) The microstructure of the 'Flash Crash': flow toxicity, liquidity crashes and the probability of informed trading. J Portf Manag 37(2): 118–128
5. Lee C, Ready M (2010) Inferring trade direction from intraday data. J Financ 46:733–746
6. Maslov S (2010) Simple model of a limit order driven market. Physica A 278:571–578
7. Mathieu P, Brandouy O (2010) A generic architecture for realistic simulations of complex financial dynamics. PAAMS 70:185–197
8. Nevmyvaka Y, Sycara K, Seppi D (2011) Electronic market making: initial investigation. http://citeseerx.ist.psu.edu/viewdoc/summary?doi=10.1.1.61.1642

Financial Forecasts Based on Analysis of Textual News Sources: Some Empirical Evidence

A. Barazzetti, F. Cecconi, and R. Mastronardi

Abstract The explosive growth of online news and the need to find the right news article quickly and efficiently cause people to adapt on events happening. The readers task is to filter out the desired information from headlines and teasers by scanning various sources formats (text, broadcasting transmission and video storage) of news articles. What people need are entities, relationships and events, which can be extracted from text by using event extraction techniques. Considering the granularity of event extraction, we present a novel approach that extracts correlation between news with a human/machine interaction. Our scope is to answer this question more efficiently: "How might stocks (e.g. Eni) react if a news is created and launched again across web news network?". This research examines a predictive machine learning approach for financial news articles analysis using a News Index Map (NIM) based on a web news decision support system for event forecasting and trading decision. Empirical evaluation on real online news data sets firstly show that only a small number of news ends up having a real impact on the security and secondly, human coding is able to extract knowledge from large amounts of data to build predictive models to provide investment decision suggestions.

A. Barazzetti (✉) · R. Mastronardi
QBT Sagl, Via E.Bossi, 4, 6830 Chiasso, Switzerland
e-mail: alessandro.barazzetti@qbt.ch; rosangela.mastronardi@qbt.ch

F. Cecconi
LABSS-ISTC-CNR, Via Palestro, 32, 00185 Roma, Italy
e-mail: federico.cecconi@istc.cnr.it

1 Introduction, Motivation and Related Literature

The World Wide Web is a vast and rapidly growing source of information. An omnipresent problem is the fact that most data is initially unstructured, i.e., the data format loosely implies its meaning and is described using natural, human understandable language, which makes the data limited in the degree in which it is machine interpretable. Bank, hedge funds and proprietary trading firms increasingly recognize the value in mining this web content to predict trading opportunities. But how does a firm harness the web's flood of unstructured data? Today, we're deluged with data. It's expanding beyond terabytes into petabytes, and even exabytes (1 million terabytes). Data in their terabytes and petabytes is providing people to comprehend a news article faster by giving them the answers to the questions like: "What happen to the next US election?" or "Which is the consequence of the civil war in Libya on financial market" or "Which companies are growing in popularity on the web/social media?" or predict some events [1, 7]. The ability to get specific answers to such questions comes with the growth of what's called big data. In the last decade, organizations have gone from just accumulating standard data to non-standard forms, from static to dynamic in a paradigm shift that's altered thinking about enterprise data assets. Innovative software has taken command to capture all this and provide new insights in organizations that are accessing big data, leading to faster, better decisions and quicker responses.

Big data is about connecting the dots of all the content that's out there by analysing a huge data set and returning a set of results in milliseconds. Thus is possible by designing an enterprise data warehouse to support technology that analyses data to help organizations enter the big data era successfully. Recently, a host of firms, including start-ups as well as established media giants, have been addressing the big data challenge, offering tools and services that mine Internet data and provide Wall Street with sentiment analysis (see [6, 8]). Because big data crunches data sets that are so large they cannot be speedily analysed by traditional database software tools, analytics is emerging with innovative software products – purposely designed for large amounts of data in all forms, including text, numbers, images and voice. In fact, sentiment analysis of natural language texts is a large and growing field. Extracting unstructured and structured data from the web pages is clearly very useful, since it enables us to pose complex queries over the data. Extracting structured data has also been recognized as an important sub-problem in information integration systems [3, 10], which integrate the data present in different web sites. While there have been several studies covering textual financial predictions, our idea is to extract automatically news of different nature (text, image, video, audio) from the web and before to semantically name the extracted data we define a-priori some correlation with human analysis.

Furthermore, in contrast to the several prototypes that predicting short-term market reactions to news, our prototype attempts to forecast medium/long trend of the major equity indices. Hence, the main goal of our study is to develop a system, named NewsMarket, based on human and machine coding, to support decision making in financial markets [2, 4]. There are at least two reasons why human input is beneficial: human brain is able to discover hidden correlations better then a machine (even if slower than a machine) and to extract hypothesis from an incomplete set of data with intuition as well as with logical deduction. On the contrary, human input is time consuming and error-prone. The NewsMarket system builds upon several fields of prior research in linguistics, textual representation, machine learning and application of itself to a problem in finance. While the NewsMarket system does not add anything new to these fields themselves, its contribution is the creation of the News Index Map (NIM) system that weaves together these disparate fields in the pursuit of solving a discrete prediction problem. More precisely NewsMarket tend firstly to establish a-priori news correlation (NIM) after tend to learn what article terms are going to have an impact on stock prices, how much of an impact they will have, and finally make an estimate of what the stock price is going into the future (medium-long term). The premise is that certain article terms such as *"tsunami"* or *"increase of unemployment rate"* will have a depressing effect on stock prices whereas other article terms like *"earnings rose"* or *"China continue to buy basic materials"* will have an increasing effect according to the fact that news represents events. Text mining approaches to predicting financial markets are comparatively rare due to the difficulty of extracting relevant information from unstructured data. Over the last decade several prototypes were developed, often without recognizing already existing systems. The first goal of this paper is to briefly describe the prototype developed by us, NewsMarket,[1] tool news of correlations detection, than can be useful for investment strategies. The second aim is to show the results of the analysis based on a dataset of 10,186 news concerning Eni[2] (9,233 news from online newspapers and 952 macroeconomic news extracting from Bloomberg) and to discuss the adequacy of text mining, especially of automated text categorization, to predict stock price movements. The full sample consists of news and 1-min closed prices of Eni from January 1, 2011 through December 31, 2011.

NewsMarket is designed to convert data information that is presented by NIM into financial markets movement tendency information, thus generating sell/buy/hold signals directly for investors. Our scope is to build a very comprehensive events[3] database, (or correlation news/events) around events that happened

---

[1] NewsMarket is a prototype created by QBT Sagl.

[2] Eni is an integrated energy company. Active in more than 70 countries, with a staff of 76,000 employees, in the oil and gas, electricity generation and sale, petrochemicals, oilfield services construction and engineering industries.

[3] Xie, Hari, and Campbell [11] write: "Events can be defined as real-world occurrences that unfold over space and time. In other words, an event has a duration, occurs in a specific place, and typically will involve certain change of state. Using this definition, "a walk on the beach", "the hurricane of 2005", and "a trip to Santa Barbara" would all qualify as events. Events are useful because they help us make sense of the world around us by helping to recollect real-world experiences (e.g., university commencement 2006), by explaining phenomena that we observe (e.g., the annual journey of migrating birds), or by assisting us in predicting future events (e.g., the outcome of a tennis match)."

in the past, events that are happening now and in the future, and make that available to investment banks and traders, professional and not. NewsMarket is able to extract semantics from news using a second component, which joins the NIM, the semantic network of concepts (SNC). SNC is, in principle, very simple. The news, as mentioned above, is broken down into strings of lemmas, eliminating the morphological structure and extracting syntactic info. The entries are associated with a synonymous database, which ensures good lexical coverage. At this point the lemmas activate the concepts within the SNC, for example lemmas "to massacre" and "Nigeria" activate the concept "Political Instability", which propagates its activation on the network, bringing with it the information on the NIM. Hence, you can obtain the activation of the above concept "Instability". The reading of the concepts activated, and their state of activation allows giving a semantic to NewsMarket's forecast. NewsMarket is a powerful tool that can be used to support decision makers in making strategic decisions. One of their objectives is to improve management judgment by fostering understanding and insights and by allowing appropriate access to relevant information. The ability to forecast the future, based only on past data, leads to strategic advantages, which may be the key to success in organizations. In real life, one would be interested not only in efforts in forecasting, but also in practical trading strategies with possibility of taking positions not only in equity markets, but in all financial asset classes. Furthermore, Tsoi [9], in their earlier studies have shown that the direction of the forecast is more important than the actual forecast itself in determining the profitability of a forecasting and trading system. Using technical and fundamental analysis, traders are usually to sell at top range and to buy at bottom range. Forecasting often plays an important role in the process of decision-making. Hence, forecasting system is always connected with the NewsMarket to improve decision-making.

We have demonstrated in our job that it is possible, with human activity, to extract knowledge from large amounts of data and to use this knowledge to build predictive models of behaviour of the market of stocks. *We have also demonstrated that only a small number of news is involved in the prediction of the behaviour of a stock, correlated and non-correlated in a similar percentage.* NewsMarket is able to help operators of trading activity in the form of alarms and trends of stock quotes [2]. According to Grimmer J. and Stewart B. M. [2], that emphasized that human coding is preferred to machine coding in to classify a subset of documents into a predetermined categorization but at the same time, differently from them we have demonstrated that only a few news with an high correlation with the equity index/event can predict more accurately and with an high probability the trend of an equity stock or an event. The rest of the paper is organized as follows. Section 2 describes our prototype, NewsMarket and their components. Section 3 provides description of the model; results are presented in Sect. 4. Section 5 delivers our conclusions and a brief discourse on future research directions.

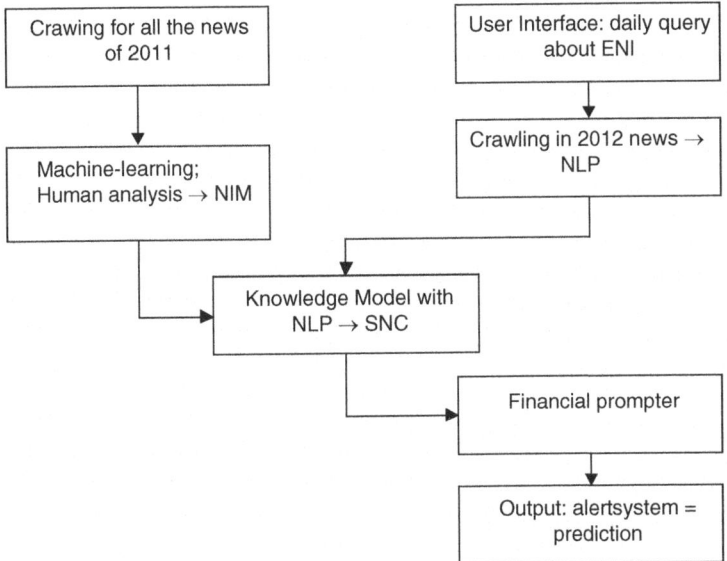

Fig. 1 NewsMarket prototype structure

2 The Prototype: NewsMarket

We developed a prototype that works on energy stock quotations. For the scope of this study we focus our attention on a specific company, Eni. The calculation of the index starts at 09:05 CET and ends with documenting prices from the ENI closing auction at 17:30 CET from Monday to Friday. We collected all the news about Eni released during all the day from Monday to Sunday. We collected the disseminated news about markets, political issues and social events with a crawling technique in database storage from different public sources from 2011 till the end of 2012. The news of the 2011 where used to generate the behavioural schema, as described, while the news of the 2012 where stored for test the trading signal of NewsMarket. The main goal of our research is to learn what news/events are going to have an impact on stock prices, how much of an impact they will have, and then make an estimate of what the stock price is going to be into the future. This prototype attempts to forecast also the 1-day behaviour for stocks and sectors. The premise is that certain terms such as "political conflicts" or "tsunami" will have a depressing effect on stock prices whereas other article terms like "earnings rose" will have an increasing effect. To properly evaluate our research questions, we designed the NewsMarket system. Figure 1 shows the main architecture of NewsMarket system and the connections among their three principal components News Index Map (NIM), Semantic Network of Concepts (SNC) and the Financial Prompter. The workflow of the system is as follows. The user queries the system asking for an investment recommendation about ENI or oil & gas sector on a daily basis of 2012. Once the query has been received, NewsMarket loads all the rules that must be checked in order to provide that recommendation within the framework model of the Semantic Network of Concepts (SNC). The SNC model is based on a previous analysis that assigned to each news/event the NIM for the necessary data. NewsMarket performs a crawl on the database to gather news about company or sector for that day of 2012: once the information has arrived, it will perform a Natural Language Process (NLP) to make interchange calls with the SNC that is in charge of assigning the NIM values. The SNC will finally write the financial reasoning ontology with all the information generated by each rule: this ontology will be sent to the selected reasoned (Financial Prompter). The result of that inference will be processed and returned to the user in an investment answer format. In what follows the main components and features of NewsMarket are depicted.

2.1 News Index Map (NIM)

News Index Map (NIM) is the first step of our research. Here, human experts analyse the online news, observe their effect on stock, and assign, if it is possible, events to categories (for example war, tsunami, change of CEO, insider trading...). Macroeconomic news will be analysed to extract a model for very short time activity, instead, all other news will be analysed to study a prediction model for investments in the medium/long term. All the news were read and analysed by human although this process is time and resource intensive. Human work is important to create an appropriate events database. More precisely, human can classify keywords with the right connotation and assign the correct polarity (bad news or good news). Sometimes, many words may have a negative connotation in one context and a positive connotation in another. For example crude oil may have a negative connotation in an article concerning "oil tanker crashed..." and a positive connotation in an earning reports. Human coders are used also to assigns weights in terms of impacts on the share price, to evaluate the timing of these effects and the correlation of the news with the referral security or stock sectors. The extent of prediction between financial and non-financial news articles and their impact on stock market prices is a complex avenue to investigate. While the information contained in financial news articles can have a visible impact on a security's price, information contained in general news that can cause a sudden price movement, is more difficult to capture and analyse. The first challenge of financial and non-financial prediction is to process a large amount of textual, audio and video information that exist for each security and sector. In an initial phase, a large number of websites were studied and the ones most suitable for the project were identified (see Table 1). For our scope, we use in this phase only Italian sources. One of the principal sources used for the analysis is *Sole24Ore.com* that offer free real-time and subscription-based services. For the news related to Eni and energy sector we use respectively Eni web site (that provides a rich linguistic structure that if properly

Table 1 Taxonomy of textual financial data

Data Source (URL)	Type	Holding period	Number	Holding period	Number
Archivio sole 24 ore	Provides news stories on company activities and breaking financial news articles	2011	8851	2012	8149
Trend-online.com	Recommendations Buy-Hold-Sell based on expert assessment	2011	24	2012	56
Borsa inside.com	Provides news concerning energy sector	2011	171	2012	271
Eni.it	Quarterly-annual reports	2011	88	2012	120
Bloomberg-trading212.com	Macroeconomic News	2011	952	2012	1200

read can indicate how the company will perform in the future) [5], *Borsainside.com and Trend-online.com* (respectively news concerning energy sector and current and past analyst recommendations for the company. This makes it possible to track the changing sentiment of analysts by following the upgrades and downgrades over time). All of them offer free news. Finally, we extract macroeconomics news from *Bloomberg database*. Information can be of three types: those derived from textual data (news article), those derived from audio and video data and those derived from numerical data (earnings, macroeconomic news,...).[4] We have gathered news articles and numerical information on a minute-by-minute basis for Eni securities. For each web site source the last news are obtained and stored in a database. The information that is retrieved from each news is the date of publication, the information source, the Url and the abstract. Abstracts constitute the corpus from which the system extracts the information. We consider firstly the abstract and the headline of each article because they usually condense the polarity of news and after we read carefully all of them to find hidden correlation. We gathered 9,234 financial news and 952 macroeconomic news articles and 256 daily or 25,873 intraday prices over the holding period. Before storing data into the database, analysts perform some pre-processing in order to extract the relevant information linked to Eni from the raw news. The pre-processing filtering is realized by human, reading news one by one. Once removed all the news that are not linked with Eni, we proceed to read again our short bucket of news and, for each of them, a team of traders assigns weights, in terms of impacts on the share price, evaluate the timing of the

---

[4]In this study we have excluded audio and video news. We will consider them for the development of the definite prototype.

effects and the correlation of the news with the referral stock sector. Hence, we identify:

1. The type of correlation (−1 means that the event is not correlated to the securities, 1 otherwise)
2. The importance of the event that is function of the time. High (H), the news have an immediate effect on securities (1-day), Medium (M), the news have an impact on the stocks during the medium term (2–30 days), Low (L) the event probably have an effect over 1 month.
3. The main objective is to classify the set of news obtained in the previous module according to their polarity: positive (G – Good), negative (N – Negative) or neutral (IN – In line).
4. Key words by selecting the article in well-defined macro categories (for example Eni, War, Macroeconomic news, ethical conflicts,...)

In this large amount of data we identify the vectors (see Sect. 2.2) of words that generate the same behavioural schema. So we can generalize the behaviour in a trend 3×3 matrix of weights: each news can be seen as a sequence of words that has a consequence for the referral stock. This consequence is the combination of the elements of the matrix, that we call NIM: it is the measure of the news and it is a scalar. Hence, we can write NIM as a function of three elements:

$$NIM = f(g(correlation); k(timing); j(effect)) \qquad (1)$$

Where g, k, j are non linear functions and can assume below value: Correlation can be $+1$ or -1; Timing $=$ H (high) $=$ 1-day; M (medium) $=$ from 2 to 30 days; L (low) over 1 month; Effect is the nature of the news/event. It can be B $=$ bad; G $=$ good; IL $=$ in line. We also discard all news articles from our dataset that could have an ambiguous effect. After discarding all the news article that are influent for our study we were left with a total of 861 financial and non-financial news article. The two macro area events returned by the NIM are: general key events and specific key events. General key events are the events that influenced indirectly the behaviour of Eni, while specific key events are the events connected directly to the security. General key events: European sovereign debt crisis, PIIGS, war, civil conflicts, political conflicts, tsunami, earthquake, and economy. Specific key events: Eni, energy, gas, oil, brent, crude oil, peer group company (BP, Shell, Saipem,...) and exploration. Starting from each macro group it is possible to create a connected graph formed by combining the events links pointing. For example, the civil war in Lybia dominates the news in 2011 and the beginning of 2011, was linked with Eni's activities in Libya (example of correlated news event). Tsunami in Tokyo, in March 2011, was a non-correlated event with Eni but it caused the worldwide financial markets' instability.

Table 2 SNC model

SNC-id		SNC-		Concepts			NIM
1	a	b	d	f	h		+4
2		f	g	j	k		−1
3		s	d	c	v	n	m n +2

2.2 Natural Language Processing and Semantic Network of Concepts

We make use of a series of software tools for natural language processing (NLP) whose aim is to help us to generate the behavioural schema of sentences related to events and to the NIM of each sentence itself: at first, we use a word segmentation tool, followed by a sense disambiguation process to identify the correct meaning of the sentences. At least we use a syntactical tool for the lemmatization of the text and a dictionary of synonymous to normalize it. Once normalized, we apply tags to the metadata that univocally identify the concept related to specific values of NIM. The NLP was applied, at first and as described, on the news of 2011 to extract the knowledge model: it is a semantic network of concepts (SNC) that consists of the association between the values of NIM with the single event related to the news. The result is a set of tagged sentences, representing events, that are united by the same effects as measured by NIM (see Table 2). Where i.e. a = "tag-african-country"; b = "tag-rebellion"; d = "tag-plant-danger". Let's try to clarify this point with an example. Imagine that the news is: *"News 325, More than 100 people are killed when religious violence flares in Mainly-Muslim towns in the north and in the southern city of Onitsha, Nigeria. with +4 NIM"*. From the point of view of semantic network, we must link the News 325 with some concept related with: (1) the fact that we are talking about an African Country (coming from Nigeria lemma), (2) and we are talking about some kind of "rebellion" and "damage to the industrial plant" (coming from Onitsha). When we link the lemma coming from News 325 (high NIM, +4) with concept into SNC, we obtain the information that also the concepts a,b and d are related with an high NIM (+4). SNC allows to disengage the NIM by the level of lemmas to that of a semantic to a higher level. Then we apply the NLP on each news of 2012 as daily queried by the user, so we obtain a metadata that is ready to be processed in the decision tree, to establish at which vector of the SNC it belongs; the NIM is attributed. The combination of the NIM from different news allows us to calculate the future trend of the stock quotes and then generate an alarm, if necessary. The future trend is calculated with an algorithm called Financial Prompter, as described in Sect. 2.3.

2.3 Financial Prompter

The Financial Prompter (FP) is an integrated event forecasting and trading decision support system. The inputs of the FP process are the daily news: they are processed through the NLP as described above to establish the correct belongings to SNC-concept in the SNC model and the corresponding NIM value. The trend is determined by the following formula where dfp means "density of future probability":

$$dfp_{te} = \sum_{i=1}^{n} nim_i c_i \qquad (2)$$

Where: te|te = 1;1 < te < = 30; te > 30 expressed in days, n is the number of news/events at time t expressed in days, nim identifies the News Index Map as described above for that news and c is the weight of each NIM. In this system, five trading rules are presented for users' choice:

1. Rule: if dfpte >> 0 => then the current trading strategy is "strong buy"
2. Rule: if dfpte > 0 => then the current trading strategy is "buy"
3. Rule: if dfpte = 0 => then the current trading strategy is "hold"
4. Rule: if dfpte < 0 => then the current trading strategy is "sell"
5. Rule: if dfpte << 0 => then the current trading strategy is "strong sell".

3 Data Analysis

To test our prototype we picked a research period of January 01, 2011 to December 31, 2012[5] to gather news articles and stock quotes.[6] We further focused our attention only on one company listed in the FtseMib,[7] that belongs to energy sector. The period chosen it gathered a comparable number of articles. We also observe that the period is characterized by different market conditions (hence different events) and would be a good test bed for our evaluation. We further limited the scope of the test on one company listed in the FtseMib. The company in question is Eni, that is included/considered in the energy sector. Thus, the data set contains a total of 256 daily or 25,873 intraday prices. The analysis differs from other work because we test our prototype especially for medium/long term financial activity with a predictive model built on human analysis of news seen as events. We feed the NewsMarket prototype with daily news since January 2012 till the beginning of 2013: each news undergoes a NLP process and introduced in the SNC model via a decision tree to obtain the correct NIM values. All the NIM values are computed as described

---

[5]Data was taken from Bloomberg, il sole 24 ore website, borsainside.com, Trend-online.com and Eni.it.

[6]Although the trading starts at 9:05 am and close at 17:30 CET we felt important to consider news article release during all the day (also news posted after closing hours) and for every days (during weekends, or on holidays).

[7]The FTSE MIB Index is the primary benchmark index for the Italian equity market.

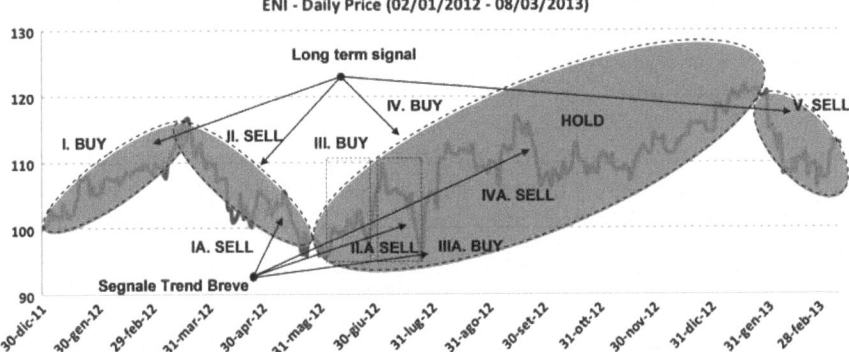

Fig. 2 Trend signal outputted from FP for Eni

above in the FP and we obtain some predictions for sell/buy signals for the chosen company for the test. Here below in Fig. 2 some exploitations of firsts results with the news that mostly occurred in determining the trend signals.

Daily impacts:

- 11/05/2012–01/06/2012: Separation Eni-Snam: reduced profit for the year
- IIA. 04/07/2012–24/07/2012: Press release Snam Sale
- IIIA. 25/07/2012: Lia release from seizure actions
- VA. 19/09/2012–28/09/2012: Corruption Nigeria

Medium term impacts:

- I. 02/01/2012–19/03/2012: Understanding of Greece, Positive Quarter, New discovery Natural Gas
- II. 20/03/2012–01/06/2012: Separation Eni-Snam; Attack on Eni plants in Nigeria, Libya Financial Irregularities
- III. 02/06/2012–03/07/2012: Acquitted Manager of Eni; Karachaganak
- IV. 25/07/2012–28/01/2013: Development Anadarko, JX Nippon Agreement; Exploratory Activities Libya; Rating Buy
- V. 28/01/2013–04/03/2013: Profit Warning Saipem, Reduce Earnings Estimates; Corruption Algeria Contracts

4 Empirical Results

We discovered that the dimension of the problem of human analysis of news and measuring their effects is possible within a reasonable timing, quantified in manpower. While measuring the impact of news we also discovered that only a small percentage of news really has importance for a single economic sector or stock market. This percentage is almost equally divided among related a non-related news but non-related news have a deeper impact in the form of events on the stock price. The predictive model, applied to the news of the year 2012 and 2013 as a back-test, is now working and the very first results encouraging us to deeply extend the time-frame of the prediction. After defining our news/events database, we would like to predict the probability, given a news article, that the trend of the stock will change. Moreover, the paper also creates a in-house news events database, in different languages, starting from a large-scale news analysis, to try to predict future behaviour. More specifically, starting from 9,234 online news, we identify a short list of them (900 circa of which 500 directly correlated with the topic) in which has been possible to identify 60 events attributable to concepts (SNC) that have significant correlation to Eni and 40 non-correlated events. We use this semantic network of concepts, that have been characterized the behaviour of Eni during 2011, to forecast the future trend. The effort in terms of human activity is measurable in 1.5 person/month for the analysis of about 10,000 news for a single sector/company. The analysis shows that training a system on Stock or Sector specific news articles led to more accurate predictions of price direction. It was reasoned that keywords specific to the company firstly and related sector secondly were more influential in determining price direction. Moreover, it was also found that training a system on specific events, correlated or not, led to a more uniform predictions of the future trend. Although the findings presented are certainly interesting, one of the limitations of this study is the small dataset. Using a larger dataset would help offset any market biases that are associated with using a compressed period of time. We will try to cover these limitations with the creation of a datasets of several years although it is a time consuming and require a large team of analyst.

5 Conclusions

To summarize, financial analysis using web data is an interesting research topic for finance industry as well as computer science. Following previous related works, this paper proves that linguistic information, with human activity, has some predictive power for finance variables like stock daily prices, and thus it can be used in the broad area of financial analysis. Hence, NewsMarket is able to help high qualified and non-qualified financial professionals with alarms or investment suggestion for medium and long term time framework. We have demonstrated that a qualitative analysis (human) can perform significantly better than a machine system. We have also made a number of improvements over previous work, and we have contributed to the literature by employing techniques not previously applied to this task. These results encourage us to explore more research topics in semantic financial analysis area.

References

1. Casti J (2012) Eventi X, Eventi estremi e il futuro della civilta'. Il Saggiatore
2. Grimmer J, Stewart BM (2013) Text as data: The Promise and Pitfalls of automatic content analysis methods for political texts. Pol Anal 21(3):267–297
3. Haas LM, Kossmann D, Wimmers EL, Yang J (1997) Optimizing queries across diverse data sources. In: Proceedings of the twenty-third international conference on very large data bases, Athens, pp 276–285
4. Jonathan C, Boyd-Graber J, Wang C, Gerrish S, Blei DM (2009) Reading tea leaves: how humans interpret topic models. In: Bengio Y, Schuurmans D, Lafferty J, Williams CKI, Culotta A (eds) Advances in neural information processing systems. MIT, Cambridge, pp 288–296
5. Kloptchenko A, Eklund T et al (2004) Combining data and text mining techniques for analysing financial reports. Intell Syst Acc Financ Manag 12(1):29–41
6. Mittermayer M (2004) Forecasting intraday stock price trends with text mining techniques. In: Hawaii international conference on system sciences, Kailua-Kona
7. Scheffer M (2009) Critical transitions in nature and society. Princeton studies in complexity. Princeton University Press, Princeton
8. Schumaker RP, Chen H (2006) Textual analysis of stock market prediction using financial news articles. In: 12th Americas conference on information systems (AMCIS-2006), Acapulco
9. Tsoi AC, Tan CNW, Lawrence S (1993) Financial time series forecasting: application of artificial neural network techniques. In: 1993 international symposium on nonlinear theory and its applications, Hawaii
10. Ullman JD (1997) Information integration using logical views. In: Proceedings of the sixth international conference on database theory, Delphi. pp 19–40
11. Xie L, Sundaram H, Campbell M (2008) Event mining in multimedia streams. Proc IEEE 96(4):623–647

Trust in a Network of Investors and Startup Entrepreneurs

Michael Roos and Anna Klabunde

Abstract We propose a simulation model of a stylized angel investor market in which the business relations are conditioned on the trust of the angel into the startup entrepreneur. Initial trust depends on social proximity between the angel and the entrepreneur and is later on updated depending on the returns an investor receives from an entrepreneur. We show that investors benefit most in terms of returns from an intermediate sensitivity of trust to returns. From an investor's perspective, too much trust keeps too many unproductive firms in the market, while too little trust leads to the termination of profitable relations because of minor productivity drops. The proportion of entrepreneurs who reach the objective of leaving the market with enough capital, however, is higher if investors lose trust quickly and stop funding early. In this sense there is a conflict of interest between investors and entrepreneurs.

1 Introduction

At least since Knight [7] it is well established that starting a new business is a decision under *true* or *immeasurable* uncertainty. An entrepreneur cannot perform any reliable quantitative risk assessment of the chances that his business will be a success, because by nature an innovation creates a new market situation for which no historical data can be available. Instead of forming probabilistic expectations, the entrepreneur exercises *judgment* on whether there are profit opportunities or not. Since there is no market for this judgment, the entrepreneur is required to start a firm. The personal traits required to exercise this judgment and to start a firm make the entrepreneur an entrepreneur.

M. Roos · A. Klabunde (✉)
Ruhr University Bochum, Universitätsstr. 150, 44801 Bochum, Germany
e-mail: michael.roos@rub.de; anna.klabunde@rub.de

In order to start a firm an entrepreneur needs capital, which he typically does not own. However, investors who could provide the necessary capital face the same immeasurable uncertainty as the entrepreneur and in addition can neither rationally assess the skills and characteristics of the entrepreneur nor his judgment. Given these problems of true uncertainty and asymmetric information and the additional problem that many young entrepreneurs cannot provide collateral for capital, it seems highly surprising that there are investors willing to give capital to startup entrepreneurs. How is it possible for entrepreneurs to raise enough capital to start and grow firms given these difficulties that seem to be sufficient conditions for market failure?

We argue that *trust* of investors in entrepreneurs is the decisive factor that makes markets for startup financing possible. Trust can be defined as "firm belief in the reliability, truth, or ability of someone or something" (Oxford Dictionaries).[1] In this view, not only the entrepreneur makes a decision based on judgment but also the investor, namely that the entrepreneur who is trusted has both good intentions and abilities making it likely that the investment will generate a positive return. Trust is then a way of forming expectations and a heuristic decision rule allowing investors to deal with the true uncertainty about the outcome of funding startups. The purpose of this paper is to analyze theoretically how the trusting behavior of investors affects how a market for startup funding works.

We propose a simple stylized model of a hypothetical angel investor market. In the model, investors trust entrepreneurs that are culturally close and form links with them. Only entrepreneurs with links to investors can obtain funding. With this funding entrepreneurs finance production in their firm and use the sales revenues to pay an interest to the investors, finance production in the next period or build up their own stock of financial capital. This decision implies the following tradeoff. Entrepreneurs would like to leave the angel investor market early and get access to other sources of funding, which suggests that they should keep a lot of their profits to themselves. But offering investors a low return decreases both their chances of receiving more funds in the future and the amount that they will get. The endogenous decision on the use of revenues changes over time due to reinforcement learning. The investors' decision on how much funding to provide to each entrepreneur in their network depends on past returns. Furthermore, investors adjust their trust in the entrepreneurs they fund. Investors are also linked to other investors and compare the return they receive from each entrepreneur with the average return of the other investors in their network. If the return from an entrepreneur is high enough, the trust in this entrepreneur increases. However, if the return is too low, the investor becomes disappointed and trusts the entrepreneur less. Below a certain minimum threshold,

---

[1] Earle [2, p. 786] distinguishes *trust* as the "willingness, in the expectations of beneficial outcomes, to make oneself vulnerable to another based on a judgment of similarity of intentions or values" from *confidence* as "the belief, based on experience or evidence (e.g., past performance), that certain future events will occur as expected". While we agree that this distinction may sometimes be quite important, it is not for our purposes. We use trust in the wider sense comprising both trust and confidence in the definitions of Earle.

the investor has no trust in a low-return entrepreneur anymore and cuts the link. Entrepreneurs that do not have links to investors and have not saved sufficiently go bankrupt and must leave the market. If the entrepreneurs have accumulated enough own capital, they leave the angel investor market voluntarily.

We simulate a calibrated version of this market over 200 periods in order to study the properties of such a market. While the calibration is inspired by some stylized facts of the U.S. angel investor market as described in [16] and [12], it is not the objective of this paper to be a realistic description of any real market in a particular time period. Such an exercise is of course relevant and interesting, but left for future research. In this paper, we focus on the basic mechanisms of our model and answer the following questions. How does the trusting behavior of the investors affect the return they can make? What is its effect on the performance of the entrepreneurs?

The literature on trust in financial markets was boosted by the recent global financial crisis (see [6, 8, 10, 11, 15]). Bottazzi et al. [1] show empirically that trust matters for the provision of venture capital and [17] and [13] find that trust is an important determinant of investment decisions in business angel finance. Closely related to our paper in terms of methodology is the paper by Gorobets and Nooteboom [5] which employs an agent-based computational model to investigate whether and under what conditions trust is viable in goods markets.

2 Trust

Trust plays a central role in our analysis. We explore how the way in which trust of investors into entrepreneurs evolves as a consequence of their interaction affects the performance of both the investors and the entrepreneurs in the market.

We assume that trust is the foundation of any business relation in our model. Without a minimum level of trust, investors are not willing to provide capital to an entrepreneur under any conditions. Investors establish connections to entrepreneurs that are "close" to them. Closeness is modeled as distance in a two-dimensional grid over which both investors and entrepreneurs are randomly scattered. While this distance could be interpreted as geographic proximity, we prefer the interpretation of cultural or social closeness. Glaeser et al. [4] and Yuki et al. [18] show that the level of trust is positively related to social closeness which is determined among other things by the number of common acquaintances, race and nationality, or organizational membership. This would mean that two people of the same nationality who are alumni of the same university trust each other more than two other from different countries and different educational backgrounds. In our model, each investor is endowed with a fixed capability of forming links to entrepreneurs, which we interpret as a time budget available to establish a personal relationship to an entrepreneur. Each investor establishes links to entrepreneurs starting with the entrepreneur to which she is closest. The further an entrepreneur is away the more costly it is for the investor to establish a connection, because it takes longer to assess the other person's personality and abilities. Connections are built until the budget for links is depleted.

The initial level of trust is the inverse of the social distance. The amount invested into the startup firm is not directly related to the level of trust, but depends proportionally on the past return the investor received from the entrepreneur. Initially, when there is no history of return decisions, all entrepreneurs receive the same amount of capital from an investor. We hence deliberately separate the decision to trust from the investment decision. Trust is a precondition for business, but the terms of the business decision depend on economic considerations about the return and not on the personal relation between the investor and the entrepreneur. However, the amount of capital provided and the level of trust are correlated because both depend in a similar way on past actions of the entrepreneur. The level of trust is updated as follows. Investors have a disappointment threshold which is a fixed percentage of the average return that other investors received from their entrepreneurs. If an investor receives a return from an entrepreneur that is below this threshold, she is disappointed and loses trust which means that the level of trust in that entrepreneur decreases by a fixed amount. Conversely, she gains more trust—again by a fixed amount—if the return received is above the disappointment threshold. In line with psychological research on trust [3,9,14], we model change of trust as asymmetric, i.e. trust increases are smaller than trust decreases. If the level of trust falls below another threshold (cutoff level), the investor does not trust the entrepreneur any longer and terminates the business relation.

We analyze the impact of different parameter values for the disappointment threshold and the size of the steps by which trust falls. These parameters describe characteristics of the investors which could be determined by socialization, personality or experience. Our main interest is how the trusting behavior of the investors affects the performance of the investors themselves, measured by the average return they can make with their investments. But we are also interested in the effects on the performance of the entrepreneurs which we measure by the proportion of voluntary exits. Entrepreneurs leave the angel investor market voluntarily if they have accumulated sufficient own capital to obtain different funding from other sources.

3 Model

There are 210 investors and 160 entrepreneurs. While there is no entry or exit of investors, entrepreneurs leave the market voluntarily or forced. To keep the number of entrepreneurs fixed, each leaving entrepreneur is replaced by a new one with a random location in the social space. All investors and all entrepreneurs are homogeneous with the exception of an idiosyncratic productivity shock each period to the productivity of each entrepreneur's firm. There is no direct interaction between entrepreneurs, but every investor is part of a constant random network with five direct ties to other investors. Investors share the information about their average return received from their entrepreneurs within their network.

In a time step of the model the following happens:

1. If investors have not exhausted their time budget on entrepreneurs, they create new links to new entrepreneurs. The cost of the links is proportional to cultural distance.
2. Investors decide with whom of their associated entrepreneurs they want to invest this period and what amount to invest with whom.
3. Investors endow entrepreneurs with capital.
4. Entrepreneurs learn their return, which is determined by a linear production function plus a stochastic component that represents the uncertainty of the environment.
5. Entrepreneurs return the investment made with them to the investors. Then they decide how much of the profit to set aside for their private wealth, how much to give to the investors, and how much to invest in the business themselves in the next period.
6. Investors receive their return from the entrepreneurs.
7. Investors update their trust towards the entrepreneurs.
8. If the trust to an entrepreneur is too low, the investor cuts the link.
9. If an entrepreneur does not receive any capital he or she goes out of business and is replaced by a random new entrepreneur.
10. If the sum of an entrepreneur's capital and private wealth is higher than his saving target he or she exits the market and is replaced by a random new entrepreneur.
11. If investors have no capital left they exit the market and are replaced by a random new investor.

The entrepreneurs employ heuristics to adapt their strategy of deciding how much of their profits to return to the investors and how much to invest themselves in the firm. First, the entrepreneur computes his profit $\pi_t = r_t - i_t$. If $\pi_t > 0$ and $\pi_t > \pi_{t-1}$, the entrepreneur seeks to do more of what he seems to have done right. First, if $\pi_t \geq p_3$, he sets an amount of size p_3 aside for his private wealth. p_3 is a parameter that is fixed for a simulation run and the same for all entrepreneurs. If $\pi_t < p_3$, he sets the full π_t aside. Then, if $p_{1,t-1} > p_{1,t-2}$, he attributes part of the increase in his profits[2] to the increase in p_1 (the amount paid as a return to the investors) and sets $p_{(1,t)} = p_{(1,t-1)} + a$, where a is the parameter for adaptation speed. If $\pi_t - p_3 < p_{1,t-1} + a$, he sets $p_{1,t} = \pi_t - p_3$. The rest of the profit, $\pi_t - p_3 - p_{1,t}$, if there is any, is distributed in the following way: If $\pi_t - p_3 - p_{1,t} \geq p_{2,t-1}$, the entrepreneurs sets $p_{2,t} = p_{2,t-1}$, where $p_{2,t}$ is the amount set aside for investment in his own business the next period. Any profit remaining is split up in half and added to $p_{1,t}$ and $p_{2,t}$ in equal proportions.

---

[2] He does not know his production function and therefore does not know the size of the stochastic component.

Table 1 Baseline parameter values

Parameter	Baseline value
Number of entrepreneurs	160
Number of investors	210
Time budget investors	10
Productivity	1.6
Variance of random component of production function	0.8
Total investment per investor and period	70
Disappointment threshold	0.6
Trust cutoff	0.2
Trust decrease	1.7
Trust increase	0.5
Adaptation speed of entrepreneurs a	5
Saving target of entrepreneurs	600
Minimum amount set aside for consumption	6
Length of run	200 steps
Size of two-dimensional grid	30 × 30 patches á 30 pixels

If $\pi_t > 0$, $\pi_t > \pi_{t-1}$ and $p_{1,t-1} < p_{1,t-2}$, he does the opposite: he increases $p_{2,t}$ in a way analogous to the one described above. If $\pi_t > 0$, but $\pi_t < \pi_{t-1}$, he increases $p_{2,t} if p_{1,t-1} > p_{1,t-2}$ in the way described above, because he believes that the lower profits are partly because $p_{1,t-1}$ was too high and $p_{2,t-1}$ was too low. Instead, he increases $p_{1,t}$ if $p_{2,t-1} > p_{2,t-2}$. If $\pi_t < 0$, $p_{1,t}$, $p_{2,t}$ and $p_{3,t}$ are all 0. In the very first year of existence, when entrepreneurs do not yet have any values to compare the current profit to, they split up equally what remains of their profit after subtracting p_3.

The model is calibrated in order to roughly match some stylized facts of the U.S. angel investor market (duration of average investment 3.5 years, proportion of investments that angels lose money on 50%, average annual rate of return 31%, distribution of returns right-skewed, average number of angel investments made by an investor per year 0.43, average number of investors per start-up 4.9) as described in [16] and [12]. The calibrated parameter values that were used as baseline values can found in Table 1.

4 Results

To assess how market performance is related to trust, we ran the model several thousand times and calculated the averages of the dependent variables over these runs in period 100 of each run. By that time the model shows stable behavior in terms of its dynamic properties.

We first discuss the effects of investors' trusting behavior on their own performance in the market measured by their average return. The trusting behavior

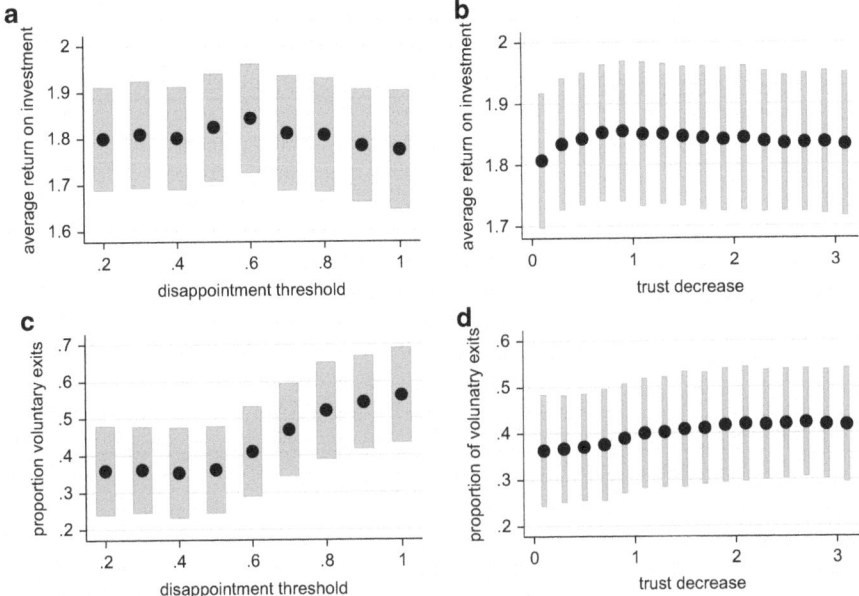

Fig. 1 Panel **a**: Effect of investors' disappointment threshold on their average return. Panel **b**: Effect of investors' decrease in trust on their average return. Panel **c**: Effect of investors' disappointment threshold on the proportion of entrepreneurs leaving the market segment voluntarily. Panel **d**: Effect of investors' decrease in trust on the proportion of entrepreneurs leaving the market segment voluntarily. All measured at step 100, average across 1,000 runs. Grey bars: mean +/− std., *black dots*: mean

is determined by two parameters of the model, the disappointment threshold and the decrease in trust after a disappointment. The disappointment threshold refers to that percentage of on individual investor's return relative to the average return of the other investors in his network below which the investor is disappointed. If, for example, the disappointment threshold is 0.8, an investor loses trust in an entrepreneur if the return from this entrepreneur is less than 80 % of what the other investors in his network receive on average. Consequently, lower disappointment thresholds mean that investors trust more.

Figure 1, Panel a, shows that the relation between the average return and the disappointment threshold is nonlinear. The average return is highest for an intermediate value of the disappointment threshold of 0.6. Both if investors become disappointed very soon or very late, their average return is lower. These effects are statistically significant as can be seen in the regression results presented in Table 2. We regressed the average return on a constant for the baseline disappointment threshold of 0.2 and a set of eight dummies for the increases in steps of 0.1. In the baseline case, the average annual rate of return is 79.9 %. The regression shows that the return at 0.6 is larger than all other average returns ($p < 0.05$) and that

Table 2 OLS regression of performance measures on disappointment threshold (Independent variables categorical, standard errors in parentheses)

Disappointment threshold	Average return	Proportion voluntary exits
0.2 (baseline)	1.7998[a] (0.0037)	0.3598[a] (0.0039)
0.3	0.0091 (0.0053)	0.0023 (0.0055)
0.4	0.0012 (0.0053)	−0.0059 (0.0055)
0.5	0.02445[a] (0.0053)	0.0023 (0.0055)
0.6	0.0437[a] (0.0046)	0.0509[a] (0.0047)
0.7	0.0112[b] (0.0052)	0.1093[a] (0.0055)
0.8	0.0073 (0.0053)	0.1609[a] (0.0055)
0.9	−0.0158[a] (0.0053)	0.1832[a] (0.0055)
1.0	−0.0252[a] (0.0053)	0.2023[a] (0.0055)
N	10,000	10,000
Adj R^2	0.0308	0.2872
Prob >F	0.000	0.000

[a] Indicate significantly different from zero at 1 %
[b] Indicate significantly different from zero at 5 %

the returns at thresholds of 0.9 and 1 are significantly lower than those at smaller thresholds.

In our model, trust decreases after every disappointment by a fixed amount. The size of these decreases in trust also affects investors' average return as visible in Fig. 1, Panel b.

The relationship is again inverted U-shaped with a maximum average return if trust decreases in steps of 0.9. If trust erodes only in small steps with a decrease by 0.1, the average return is smallest at 1.807. Slightly larger and much larger steps generate about the same average return, which is about 0.025 higher than in the baseline. A statistical assessment of the effects can be found in Table 3.

So far, we focused on the effects of one trust parameter keeping the other one fixed. With nonlinear relationships it is not straightforward to predict how the dependent variables change if both independent variables are changed. In Fig. 2, we show how the investors' average rate of return depends on the decrease in trust and the disappointment threshold. The figure shows clearly that the rate of return is highest if the decrease in trust is low and the disappointment threshold is high. This means that investors maximize their income if they lose trust early, but in small steps. Given the hump-shaped effects of the variables in isolation, this result is somewhat surprising and indicates the presence of interaction effects.

Summing up, the return received by investors is a hump-shaped function of both the disappointment threshold and the size of trust reductions. This means that investors benefit most from an intermediate sensitivity of trust to the behavior of the entrepreneurs. If they maintain a high level of trust, either because they become disappointed only when they receive very low returns or because they lose trust in very small steps, the returns they receive are relatively low. The same is true if they are quickly disappointed or decrease their trust level strongly after a disappointment.

Table 3 OLS regression of performance measures on size of decreases in trust (Independent variables categorical, standard errors in parentheses)

Trust decrease	Average return	Proportion voluntary exits
0.1 (baseline)	1.8072[a] (0.0036)	0.3632[a] (0.0038)
0.3	0.0264[a] (0.0051)	0.0038 (0.0053)
0.5	0.0351[a] (0.0051)	0.0075 (0.0053)
0.7	0.0451[a] (0.0051)	0.0126[b] (0.0053)
0.9	0.0479[a] (0.0051)	0.0262[a] (0.0053)
1.1	0.0430[a] (0.0051)	0.0373[a] (0.0053)
1.3	0.0431[a] (0.0051)	0.0393[a] (0.0053)
1.5	0.0390[a] (0.0051)	0.0455[a] (0.0053)
1.7	0.0363[a] (0.0044)	0.0476[a] (0.0046)
1.9	0.0341[a] (0.0051)	0.0539[a] (0.0053)
2.1	0.0366[a] (0.0051)	0.0566[a] (0.0053)
2.3	0.0312[a] (0.0051)	0.0553[a] (0.0053)
2.5	0.0279[a] (0.0051)	0.0575[a] (0.0053)
2.7	0.0293[a] (0.0051)	0.0599[a] (0.0053)
2.9	0.0298[a] (0.0051)	0.0559[a] (0.0053)
3.1	0.0260[a] (0.0051)	0.0546[a] (0.0053)
N	16,999	16,999
Adj R^2	0.0076	0.0269
Prob >F	0.000	0.000

[a]Indicate significantly different from zero at 1 %
[b]Indicate significantly different from zero at 5 %

For the proportion of voluntary exists of entrepreneurs the picture is different. The share of entrepreneurs leaving the angel market segment voluntarily increases if investors lose trust more quickly, i.e. with higher disappointment thresholds (see Fig. 1, Panel c) and larger trust decreases (Fig. 1, Panel d). For high levels of trust, the proportion of voluntary exists is low and most exits are forced because the entrepreneurs go bankrupt.

The regressions in Tables 2 and 3 (columns on the right) show that in the baseline case with a low disappointment threshold of 0.2 and small decreases in trust of 0.1 about 35 % of the entrepreneurs leave the angel investor market voluntarily because they have accumulated sufficient funds to have access to other sources of funding. This proportion increases significantly if the disappointment threshold exceeds 0.5 and reaches a maximum of 56 % for the highest disappointment threshold of 1. The positive effect of the size of trust decreases starts earlier and evolves more gradually. The maximum proportion of voluntary exits seems to be reached at a decrease per step of 2.7, but the difference to the neighboring categories is not statistically significant at conventional levels.

As in the case of the investors' average return, we look at the joint effects of the two trust variables on the proportion of voluntary exits. Figure 3 shows that higher levels of both variables increase the number of exits, but the effect of

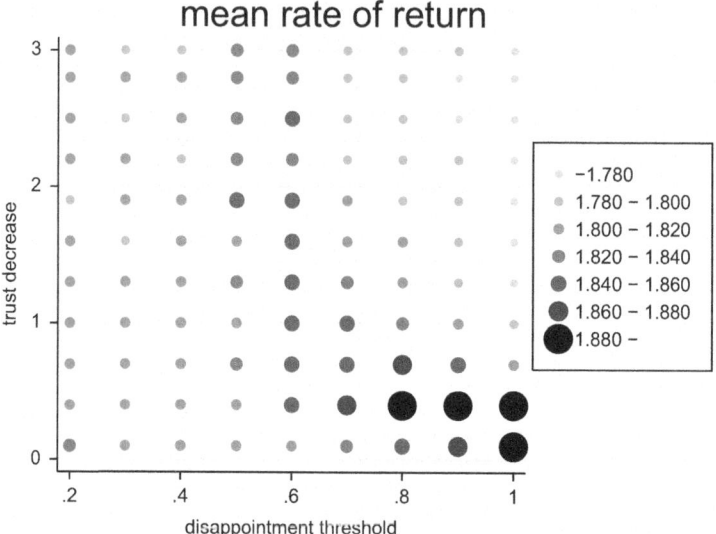

Fig. 2 Average return of investors as a function of the disappointment threshold and the decrease in trust, measured at step 100, average across 300 runs per parameter combination

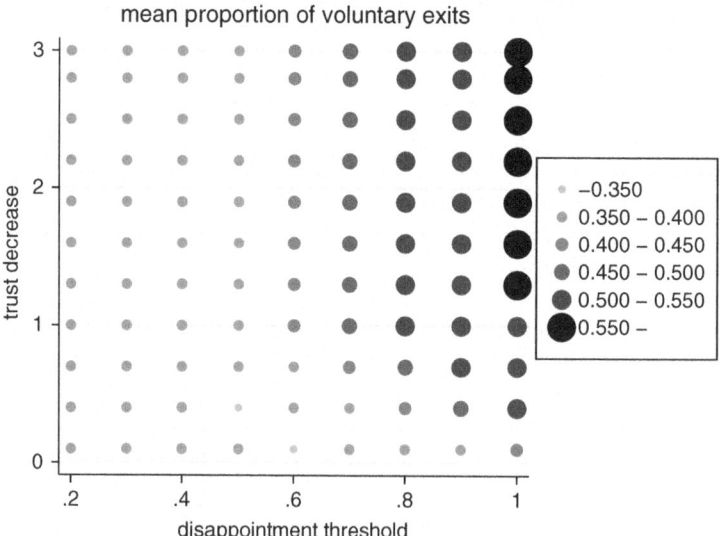

Fig. 3 Average proportion of voluntary exits as a function of the disappointment threshold and the decrease in trust, measured at step 100, average across 300 runs per parameter combination

the disappointment threshold is stronger than the effect of the decrease in trust. Given the results from the separate analyses of variables, this finding is not surprising. In Fig. 1, Panels c and d it is visible that both trust variables affect the voluntary exits positively and that the effect of the disappointment threshold is the larger of the two.

That less tolerant trusting behavior by investors is beneficial for entrepreneurs in the sense that a lower proportion is forced to leave the market involuntarily, is somewhat counterintuitive. One might have expected that more entrepreneurs can attain the savings target that allows them to leave the market voluntarily if investors trust them more. In order to understand the results, it is important to remember that trust affects the funding decision of investors in a non-linear way. As long as trust is present, it has no impact on the amount of capital transferred to the entrepreneur. Only if trust has fallen below the cutoff threshold, the investor suddenly stops the transfer of funds. This implies that the effect of trust on the behavior of entrepreneurs (voluntary exits and decision on how much interest to pay to the investor) is a population effect rather than an individual learning effect. If trust vanishes only slowly, many entrepreneurs with large negative productivity shocks remain in the market for a long time. This means that those entrepreneurs have low earnings so that their ability both to accumulate own capital and to return capital to the investors is low. This leads to low returns of the investors and a low proportion of voluntary exists. On the other hand, if investors cut relations quickly (high disappointment threshold and high trust decreases), even small productivity drops can end connections between investors and entrepreneurs. The effects are mainly negative for the investors as they forgo returns from entrepreneurs that are in principle well positioned and willing to return a lot. These investors will quickly find other investors and also increase their own capital stock so that they have good chances to reach the savings target.

From the perspective of investors it is best if they are easily disappointed but lose their trust only slowly. For entrepreneurs, high levels of the disappointment threshold are favorable as well, but they would also benefit from large decreases in trust. To a certain extent there is a conflict of interests between investors and entrepreneurs about the optimal trusting behavior.

5 Conclusions

We propose a simulation model of a stylized angel investor market in which the business relations are conditioned on the trust of the angel into the startup entrepreneur. Initial trust depends on social proximity between the angel and the entrepreneur and is later on updated depending on the returns an investor receives from an entrepreneur. While trust is sufficiently high to maintain a business relation, it has no impact on the level of investment which is proportional to the returns from an entrepreneur. The lack of trust, however, terminates the relation. Entrepreneurs use the fund from the investors to finance production in each period which is subject to idiosyncratic productivity shocks. Entrepreneurs' implicit objective is to leave the angel investor market, for which they need sufficient own capital. They accumulate capital from the profits. They must decide to allocate profits on accumulating own capital, reinvesting into the firm, and returning capital to the investors. In our model there is an indirect role of trust for market performance, as investment decisions do not depend directly on the level of trust, but only on whether there is sufficient trust of the investor in the entrepreneur or not. Yet investors as a group can influence their average rate of return by changing their trusting behavior through population effects in the market. We show that investors benefit most in terms of returns from an intermediate sensitivity of trust to returns in the sense that trust erodes quite early, but in small steps. From an investor's perspective, too much trust keeps too many unproductive firms in the market, while too little trust leads to the termination of profitable relations because of minor productivity drops. The proportion of entrepreneurs who reach the objective of leaving the market with enough capital, however, is higher if investors lose trust quickly. In this sense there is a conflict of interest between investors and entrepreneurs.

References

1. Bottazzi L, Da Rin M, Hellmann TF (2011) The Importance of trust for investment: evidence from venture capital. NBER working paper series 16923. http://www.nber.org/papers/w16923. Cited4Feb2013
2. Earle T (2009) Trust, confidence, and the 2008 global financial crisis. Risk Anal 29:785–792
3. Eiser JR, White MP (2005) A psychological approach to understanding how trust is built and lost in the context of risk. University of Sheffield. http://www.kent.ac.uk/scarr/events/trustcontext.htm
4. Glaeser EL, Laibson DI, Scheinkman JA et al (2000) Measuring trust. Q J Econ 115:811–846
5. Gorobets A, Nooteboom B (2006) Agent based modeling of trust between firms in markets. In: Bruun C (ed) Advances in artificial economics. Lecture notes in economics and mathematical systems. Springer, Berlin
6. Guiso L, Sapienza P, Zingales L (2008) Trusting the stock market. J Financ 63:2557–2600
7. Knight F (1921) Risk, uncertainty, and profit. Hart, Schaffner & Marx/Houghton Mifflin Co., Boston
8. Mayer C (2008) Trust in financial markets. Eur Financ Manag 14:617–632
9. Poortinga W, Pidgeon NF (2004) Trust, the asymmetry principle and the role of prior beliefs. Risk Anal 24:1475–1486
10. Sapienza P, Zingales L (2012) A trust crisis. Int Rev Financ 12:123–131
11. Sapienza P, Toldra A, Zingales L (2007) Understanding trust. NBER working paper series 13387. http://www.nber.org/papers/w13387.Cited4Feb2013
12. Shane S (2012) The importance of angel investing in financing the growth of entrepreneurial ventures. Q J Financ. doi:10.1142/S2010139212500097
13. Sudek R (2006) Angel investment criteria. J Small Bus Strategy 17:89–103
14. Taylor SE (1991) Asymmetrical effects of positive and negative events: the mobilization-minimization hypothesis. Psychol Bull 110:67–85
15. Tonkiss F (2009) Trust, confidence and economic crisis. Intereconomics 44:196–202

16. Wiltbank R, Boeker W (2007) Returns to angel investors in groups. Ewing Marion Kauffman Foundation & Angel Capital Education Foundation. Available via SSRN. http://papers.ssrn.com/sol3/papers.cfm?abstract_id=1028592.Cited4Feb2013
17. Wong A, Bhatia M, Freeman Z (2009) Angel finance: the other venture capital. Strategy Change 18:221–230
18. Yuki M, Maddux WW, Brewer MB et al (2005) Cross-cultural differences in relationship- and group-based trust. Pers Soc Psychol B 31:48–62

Banks and Their Contagion Potential: How Stable Is Banking System?

Mitja Steinbacher, Matjaz Steinbacher, and Matej Steinbacher

Abstract We measure contagion potential and stability of banking system on a randomized version of the credit contagion model by Steinbacher M, Steinbacher M, Steinbacher M (2012) Credit contagion in financial markets: a network-based approach. Available via SSRN. http://papers.ssrn.com/sol3/papers.cfm?abstract_id= 2068716. Cited 30 Jan 2013. We introduce two estimators of the contagion potential of banks (liquidity-loss potential and α-criticality index (Steinbacher M, Steinbacher M, Steinbacher M (2012) Credit contagion in financial markets: a network-based approach. Available via SSRN. http://papers.ssrn.com/sol3/papers. cfm?abstract_id=2068716. Cited 30 Jan 2013)) and introduce Shannon's entropy as a stability estimator. Our approach is systemic in that it enables an overall estimation of the capacity of the banking system to provide liquidity. Mechanism developed can be employed for measuring systemic risk of banking system as a whole.

1 Introduction

Schweitzer et al. [15] acknowledge that

> We need an approach that stresses the systemic complexity [...] that can be used to revise and extend established paradigms in economic theory.

M. Steinbacher (✉)
Faculty of Business Studies, Catholic Institute, Ljubljana, Slovenia
e-mail: mitja.steinbacher@gmail.com

M. Steinbacher
Kiel Institute for The World Economy, Kiel, Germany
e-mail: matjaz.steinbacher@gmail.com

M. Steinbacher
University of Donja Gorica, Podgorica, Montenegro
e-mail: matej.steinbacher@gmail.com

It would need a special anthology to acknowledge all the authors that have contributed to the knowledge about dynamics and complexities of current financial system. Specifically, we focus on structural characteristics of banking system and on credit default contagion by the use of artificial network approach in order to build a mechanism for identifying and measuring the stability of the banking system relative to various degrees and types of mutual exposures of banks. The mechanism is tested on credit contagion model [16], who model idiosyncratic and systemic shocks and their propagation through a banking system. The model is based on credit events that spread along the banking system and subsequently influence the behavior of the system; credit events are exogenous to simulator and are spurred either by financial markets, business sector, government, or private individuals for whatever reason.

Banking system is a subsystem of a financial system. The latter resembles a network of interconnected financial and non-financial agents, whose decisions are likely to produce a rich and unprecedented structure; see [1] for a reference. Allen et al. [3] provide an extensive list of references to the research conducted on various aspects of a financial system and credit contagion, focusing basically on particular applications. Allen and Gale [2] further study banking system and its responds to contagion in various network structures based on mutual deposits of banks, which is partly similar to a setup we use.[1] Eisenberg and Noe [8] for instance develop an algorithm of a natural measure of systemic risk based on waves of defaults needed to induce failure, while Upper and Worms [17] perform a contagion test on German banking system and show that the failure of a single bank could have led to the collapse of up to 15 % of overall banking assets.[2] Some [1, 12, 15] studies show that banking system is characteristic for a few large banks that are linked with many smaller banks.

2 The Model

The model is taken from [16]. It consists of a finite set of 40 banks; 14 are big banks each having over $900 billion in assets, 17 are medium with more than $100 billion and less than $700 billion in assets, and the rest are small with less

---

[1]Consult also the studies by Freixas, Parigi and Rochet [9] about banks under uncertainty of withdrawals, where banks are connected through interbank credits, the desing of financial networks that minimize the trade-off between risk sharing and the potential for collapse presented in [14] and Dasgupta's [6] study about banks' crossholdings of deposits as a source of contagion. Furthermore, reader shall also consult de Vries [7] and his dependency between banks' portfolios of assets and potential for systemic breakdown, Haldane and May's [11] study of contagion in financial markets, Gai and Kapadia's [10] model of contagion in financial networks, Cifuentes et al. [5] model of financial institutions that are connected via portfolio holdings, and the study of Jorion and Zhang [13], who show credit contagion via counterparty effects.

[2]See [4] for stress test on Austrian interbank network structure with respect to the default of a single bank.

than $100 billion in assets. A cumulative initial value of assets in the network is $25,951.16 billion and it is the same for all network topologies used in simulations. Initial capital of all banks from the model is taken as of December 31, 2011 from banks' annual reports. Mutual exposures of banks as well as their mortgage lending are distributed among banks at random and once determined remain the same in all initial periods of all banking network topologies used.

The banking system is treated as a complex system built by semi-autonomous agents that make decisions on their own behalf and follow certain generally imposed rules. Key assumptions of the model are:

1. Liquidity can either be in passive or active mode;
2. Connections among banks represent liquidity flows;
3. Initial default risk of banks is arbitrary and bank-specific;
4. Liquidity is defined in terms of a monetary unit of value.
5. Banks are not allowed to change their strategies (no autonomous change in portfolios, no autonomous change in lending, no recapitalization, etc.)
6. There is no lender of last resort.
7. Each network exists for T time units.

2.1 Banks

The dynamics of assets of bank i can be viewed as an ordered sequence $\{a_{i,t}\}_{t=0}^{T}$ that develops in time t as shown by (1).

$$a_{i,t} = h_{i,t} + b_{i,t} + n_{i,t} + \sum_{j \in \mathcal{E}(i)} q_{ij,t}. \tag{1}$$

Here $a_{i,t}, h_{i,t}, b_{i,t}$ and $n_{i,t}$ denote values of bank i total assets, mortgage loans, bonds and non-trading assets in time t, while $q_{ij,t}$ denotes its holdings of interbank assets with bank j at the same time. $\mathcal{E}(i)$ is the subset of all liquidity flows from i to j; the sum operator in (1) goes over all banks j in which i holds a portion of its interbank holdings.

The value of capital of bank i develops according to its profits and losses in time (Table 1). The dynamics can be depicted as

$$c_{i,t+1} = c_{i,t} + p_{i,t} \tag{2}$$

$$p_{i,t} = a_{i,t} - a_{i,t-1} \tag{3}$$

As for the banking network a standard notation from graph theory applies. $\mathcal{G}$ is a directed random graph that consists of the nodes of the set $\mathcal{V} \subseteq \mathbb{Z}$ and let $\mathcal{E} \subseteq [\mathcal{V}]^2$ include all connections in $\mathcal{G}$; a connection is a two-element pair $(ij) \in \mathcal{E}$.

Table 1 Banks' balance sheet

Assets	Liabilities
Interbank assets (non-tradable)	Equity
Loans to non-banking entities	Interbank liabilities
Tradable assets	Other debt
Other non-tradable assets	

Connections represent the intensity of flows passing from node i to node $j : j \neq i \mid \{j, i\} \in \mathcal{V}$ and vice versa $(i : i \neq j \mid \{i, j\} \in \mathcal{V})$. Every connection is ascribed a non-negative real number k that is defined by the transformation $f : [\mathcal{V}]^2 \to k \in \{\Re^+ \wedge 0\}$. In our case k is expressed in monetary units and represents interbank liquidity. An overall liquidity in time t is a sum of flows over all connections in the network in time t:

$$K_t = \sum_{|E|_t} f(\cdot, \cdot)_t \tag{4}$$

By assumption banks are not allowed to recapitalize losses and in all cases they go default when the level of capital they hold in their balance sheets falls short of the obligatory Tier 1 capital in total assets (4 %). Moreover, a default of any bank from the banking system affects the whole system through mutual connections of banks with the defaulted bank.

Say, bank j is a debtor of bank i (i.e. $(ij) \in [\mathcal{V}]^2$) and denote its debt as q_{ij}. Now, let $\mathcal{F} \in \mathcal{V}$ be a non-empty set of all defaulted banks from the banking system $\mathcal{G}$ and let bank j go bankrupt (i.e. $j \in \mathcal{F}$) at time t. Bank j deteriorates balance sheet of i for the amount of wd_i that is now under pressure to finance those write-downs. For the sake of financing those losses bank i uses capital; in this case capital acts as a protection buffer from default. Its stock of capital evolves as

$$c_{i,t+1} = c_{i,t} + p_{i,t} - wd_i \tag{5}$$

$$wd_i = (1 - rr_i) \cdot q_{ij,t} \tag{6}$$

Here (6) shows an immediate write-down of wd from the balance sheet of bank i after bank j went bankrupt, while $rr_i = (0, 1)$ are partial recoveries; recovery rates are exogenous and randomly distributed among banks. In principle, the stronger the exposure against any defaulted debtor and the lower the reserved capital buffer of the respective creditors, the greater the potential for credit contagion.

We have thus seen how a unit of wd_i can cause further losses. This spill-over depends on the importance of bank that was hit by a liquidity shock. We will use liquidity loss-potential estimator LLP_i and bank specific α-*criticality* index as measures of influence; see Table 2 for the definitions.

Table 2 Influence estimators used in simulations

Influence estimator	Definition
Liquidity-loss potential	$LLP_i = 1 - \frac{\int_0^{e^{(K_{i,T}/\kappa_i)}} e^\lambda \, d\lambda}{e-1}$
α-criticality	$\alpha_i = \frac{\sum_{j=1}^{|\mathcal{V}|} a_{j,T} + a_{i,t}^*}{\sum_{k=1}^{|\mathcal{V}|} a_{k,t}} \mid i \in \mathcal{F} \wedge \{j,k\} \in \mathcal{V} : j \neq i \Rightarrow$
	$\Rightarrow A_i = \frac{\int_0^{e^{\alpha_i}} e^\lambda \, d\lambda}{e-1}$

$K_{i,T}$ from the table is time T level of liquidity in the network after bank i went bankrupt in period t, as given by (4), while $a_{j,T}$ is time T level of assets in the network (1) and $a_{i,t}^*$ are assets of bank i just before it got bankrupt in time t.

Liquidity loss potential is more volatile than α-criticality index and less indicative of changes in network structure; the latter responds to actual defaults of banks, the former measures changes in liquidity that are not necessarily preceeded by defaults. Potentially useful is also a $[E(A)/E(LLP)]_r$ ratio that measures a rate at which a unit of liquidity lost in banking system r leads to a subsequent loss of a unit of its assets (i.e. both are under expectations' operator). In measuring capacity of banking system to provide liquidity we will make use of a composite influence estimator

$$CIE_r = \left[\sqrt{E(A) \cdot E(LLP)}\right]_r \qquad (7)$$

Say, each connection of the respective system r carries BI_r information about the default of certain portion of liquidity (or assets, respectively). Now, set composite influence estimator as a proxy for this influence and define BI_r as

$$BI_r = [CIE/S(\mathcal{V})]_r \qquad (8)$$

2.2 Interbank Connections and Risk

By construction every bank i is given a subset of outgoing links $\mathcal{L}(i) : |\mathcal{L}(i)| = l_i \mid l_i \sim U(0, \|\beta \times (|\mathcal{V}|-1)\|); \beta = [0,1]$ to banks debtors from the set of all debtors $d \in \mathcal{D}$, such that $(id) \in [\mathcal{V}]^2$.

Let each debtor be ascribed a non-negative probability $p_d = (0,1)$ of defaulting on its debt and let p be derived from logistic function

$$p(y) = \frac{1}{1+e^{-y}}, \qquad (9)$$

where $y = f(x_1, x_2, \ldots, x_n)$ is a function of specific factors x_i from a set of potential factors that are influencing the behavior of d. Typically $p(y)$ would be

Table 3 Risk profiles of banks

Risk group (y_i)	Share in the population of banks(α_i)
$y_1 \leq -5$	α_1
$-5 < y_2 \leq -2.5$	α_2
$-2.5 < y_3 \leq 0$	α_3
$0 < y_4 \leq 2.5$	α_4
$2.5 < y_5 \leq 5$	α_5
$y_6 \geq 5$	α_6
	s.t. $\sum_i \alpha_i = 1$

obtained by maximum likelihood regression on the logit, the inverse of the logistic function. It is not our aim to analyze the structure of risk profiles as such. We use simple heuristics instead and assign y to each debtor d by a neutral mechanism based on Table 3. Probability to default on its obligation is then calculated from (9). Defaults of those debtors are independent and shall be drawn as random variables.

In case any debtor defaults, $(1 - rr_d)$ share of its liquidity leaves the banking network. Creditors of defaulted debtors need to recover those losses by their capital, respectively. This capital turns liquid and becomes available for liquidity needs of the banking network. If there is sufficient capital held with banks to withstand the losses, the network will suffer no net outflow of liquidity and vice versa.

Now, let the losses of liquidity caused by the defaulted debtor d to bank i be denoted by LL_{id} and let $p(y_{id})$ denote a probability that debtor d from risk group y defaults on its debt to creditor i. As risk profiles of banks are independent by assumption, a probability that the banking network looses at least a unit of liquidity is given as

$$max \ \{p(y_{id})\} \tag{10}$$

s.t.

$$d \in \mathcal{D} \ \wedge \ (id) \in [\mathcal{V}]^2 \tag{11}$$

This loss of liquidity causes certain instability to the banking network and increases its vulnerability against further liquidity losses; in case liquidity loss is large enough it induces serious structural changes (i.e. credit contagion).

2.3 Stability of Banking System

Each unit of liquidity that is released to the interbank by creditor i to debtor d can be seen as a moving particle x_i with strictly positive probability of leaving the network. Say that X_i is a variable, defined on a sample space $\mathcal{S} = \{0, 1\}$. Say further that $x = 0$ means that particle stays in the network and $x = 1$ that it leaves. Then the probability mass function for X_i is defined as

$$P(x_s)_i = \Pr(X_i = x) = \Pr(\{s \in \mathcal{S} : X_i(s) = x\}) \qquad (12)$$

Creditor i has $|\mathcal{L}(i)|$ debtors d. Should any debtor d default on its debts to creditor i, all of its connections cease to exist and this liquidity is lost[3] (e.g. they share the bank's risk). Now, let $p(x_s = 1) = p(y_{id})$ denote the probability that bank d goes default and let **p** hold probabilities p_i for every bank to go default. Hence, we can apply Shannon's information entropy (13) to **p** and obtain an array **p*** of available bits of information about the defaults for all connections from the banking network.

$$H(X)_i = -\sum_s [P(x_s) \cdot \log_2 P(x_s)]_i \qquad (13)$$

The higher the $H(X)_i$ the higher uncertainty in predicting the default of bank i. Distribution function (13) has four turning points important to our analysis:

$$p_1^* : min\,[H(p_1^*)'/H(p_1^*)''],$$
$$p_2^* : H(p_2^*)' = 1,$$
$$p_3^* : H(p_3^*)' = -1,$$
$$p_4^* : max\,[H(p_4^*)'/H(p_4^*)'']$$

Based on those turning points and two rules of thumb are seven stability domains as summarized in Table 4.[4] Vector **p*** carries liquidity-loss potentials and α-criticality of all connections from the network. Stability index of the banking system $\mathcal{V}_r$ is obtained as a weighted average of all values in $(\mathbf{p}^*)_r$.

$$S(\mathcal{V})_r = (\mathbf{w}^T \cdot \mathbf{p}^*)_r \qquad (14)$$

where **w** is a vector of weights.

Ranks of stability and contagion potential of the banking systems are based on the following corollaries:

1. The higher the stability index, the less stable the system, expected contagion potential unchanged.
2. The higher the expected contagion potential, the weaker the system in terms of providing liquidity, stability unchanged.
3. Banks from safe and stable domain are not contagious.
4. Contagion potential and stability are independent.

---

[3] Recoveries are kept on creditors' balance sheets and do not enter the interbank lending market.
[4] 37 banks are in the 1st domain with two bordering on the 2nd; one bank is in the 6th domain and it has relatively weak liquidity loss potential and low α-criticality index.

Table 4 Stability domains

Domain	Conditions	Description of stability domain
1.	$P(y_{id}) = [0, 0.025) \Rightarrow$ $\Rightarrow H[P(y_{id})] = [0, 0.169)$	*Safe and stable. Limited uncertainty. Eventual default would come as a surprise as it would carry at most 0.169 bits of unknown information.*
2.	$P(y_{id}) = [0.025, 0.176) \Rightarrow$ $\Rightarrow H[P(y_{id})] =$ $[0.169, 0.671)$	*Relatively less safe and less stable domain than domain 1. Domain with relatively large volatiliy in terms of risk profiles and the quality of information available about banks. Defaults come at lower surprise than in the previous domain.*
3.	$P(y_{id}) = [0.176, {}^1/_3) \Rightarrow$ $\Rightarrow H[P(y_{id})] =$ $[0.671, 0.918)$	*Stabilized safety at yet lower levels. The domain is more homogeneous than the previous one. Some information (at most 0.329 bits) is available about the default risk of respective banks. Some banks from the group might actually default.*
4.	$P(y_{id}) = [{}^1/_3, {}^2/_3) \Rightarrow$ $\Rightarrow H[P(y_{id})] = [0.918, 1]$	*The most unpredictable domain. The least information is available about the risk profiles of banks; we effectively dispose off only at most 0.082 bits.*
5.	$P(y_{id}) = [{}^2/_3, 0.823) \Rightarrow$ $\Rightarrow H[P(y_{id})] =$ $[0.671, 0.918)$	*Stabilized to some degree and yet less safe than previous domain. The domain is also less homogeneous than the previous one. Information (at most 0.329 bits) that is available about the default risk of respective banks is negative. The majority of banks from the group might actually default.*
6.	$P(y_{id}) = [0.823, 0.975) \Rightarrow$ $\Rightarrow H[P(y_{id})] =$ $[0.169, 0.671)$	*Unsafe and on average relatively stable domain. The domain shows large volatiliy in terms of safety and the quality of information available. Survival comes at higher surprise than in the previous domain.*
7.	$P(y_{id}) = [0.975, 1) \Rightarrow$ $\Rightarrow H[P(y_{id})] = [0, 0.169)$	*Unsafe and stable. Limited uncertainty. Eventual survival would come as a surprise as it would carry at most 0.169 bits of unknown information.*

3 Results

We tested the stability of 100 independent banking systems consisting of 40 banks. As for the risk, the majority of banks belong to the safe and stable domain. Nine banks are somewhat less predictive and bank 18 is unpredictive; it carries

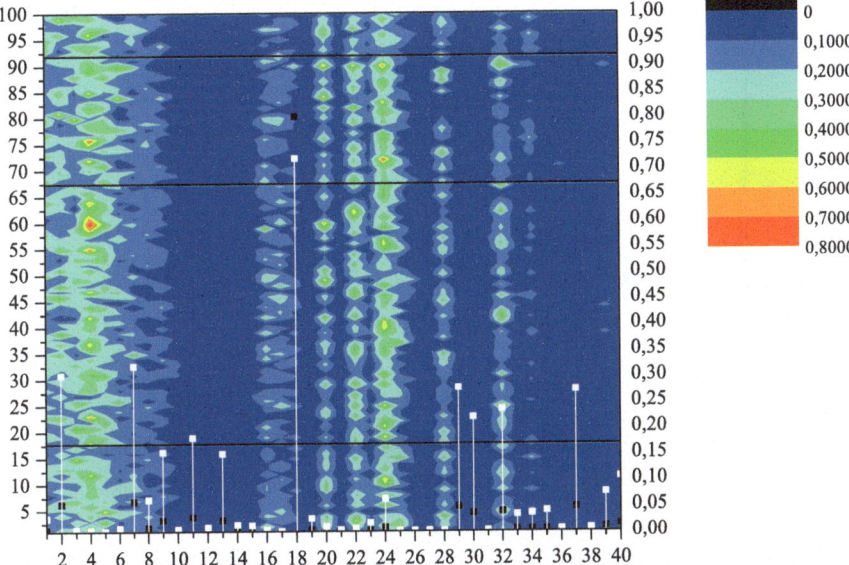

Fig. 1 Liquidity loss potential (LLP_i) of individual banks in respective topologies. Banks are indexed on *bottom axis* and topologies on the left. *Right axis* shows expected probability of default (*black dots*) and Shannon's entropy (*white dots*) for the banks as derived from (13)

low to moderate power to disrupt the banking system and spur contagion. On the other hand, bank 4 carries the strongest power to disrupt the banking system and cause contagion, but it is well within the safe and stable domain (i.e. tenth safest). Capacity of banking system to provide liquidity and remain stable thus depends essentially on the influence and risk profiles of the least stable group of banks.[5]

Banking system 18 shows the lowest overall stability ($S[\mathcal{V}_{18}] = 22.756$); 76% of the score is due to banks from domains 3–7 and 28% of the score is due to its 9 connections with bank 18. The system has 236 connections, making it ninth most uncertain on average. Excluding safe and stable banks from the score makes the system 52 the least stable ($S(\mathcal{V}_{52}|\ p_i > 0.025) = 17.38$) and followed by system 18 ($S[\mathcal{V}_{18}|\ p_i > 0.025] = 17.283$). Figures 1–6 in the sequel confirm a relatively large potential for disrupting the banking system of some banks. The last two suggest to remove banks from the safe and stabe domain when ranking banking systems' contagion potential relative to their stability.

---

[5]See Figs. 7 and 8 for expected stability index and the contributions to the overall stability by stability domains. Figures suggest elimination of the safe and stable domain from further comparative analysis of stability and contagious potential of banking systems.

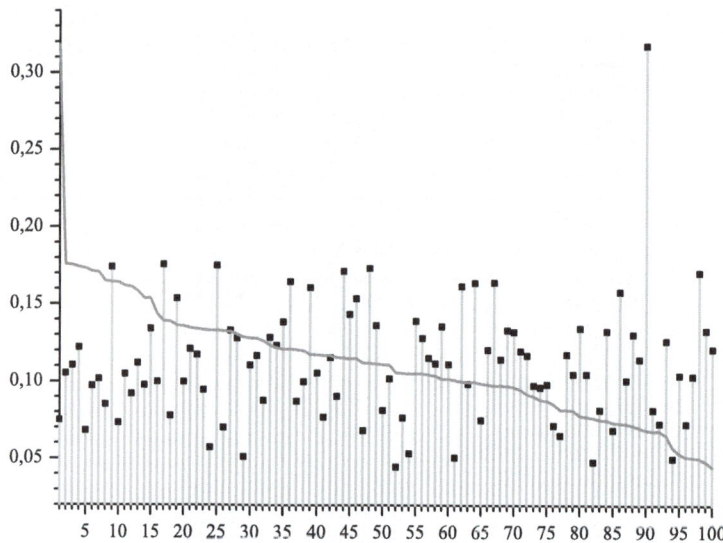

Fig. 2 Expected liquidity loss potential of respective banking system. A line is an ordered sequence

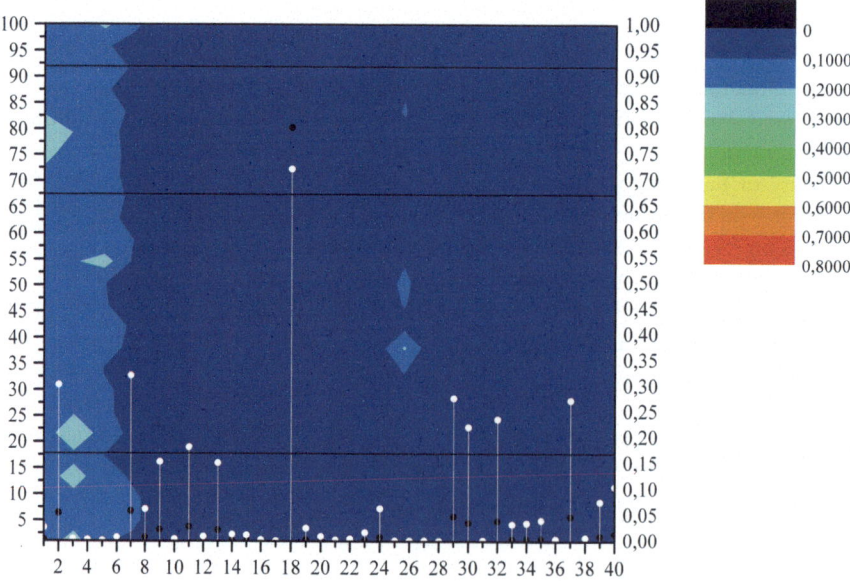

Fig. 3 α-criticality of individual banks. Scaling, topologies and indexation are the same as in Fig. 1

Banks and Their Contagion Potential: How Stable Is Banking System? 171

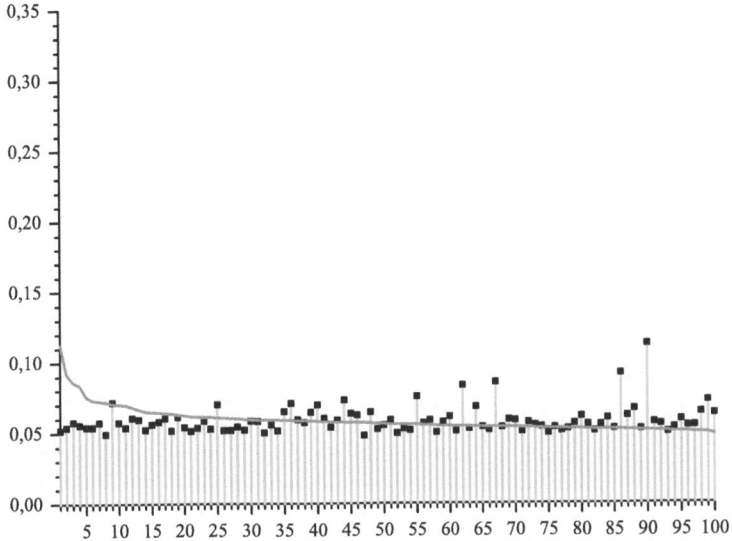

Fig. 4 Expected α-criticality of respective banking system. The same scales as in Fig. 2. A line is an ordered (in descending order) sequence

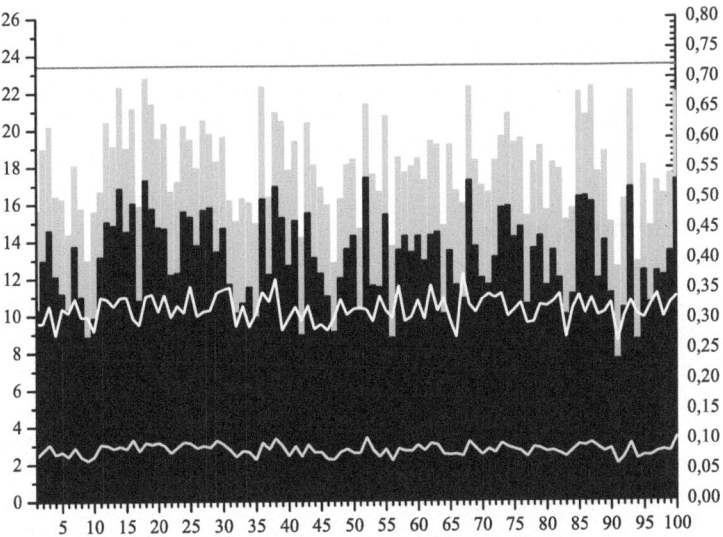

Fig. 5 A structure of stability index (columns; *left axis*). *Black* is a contribution of banks from domains 3–7; *gray* is the contribution of banks from domain 2 and *light gray* is the contribution of banks from domain 1 to the overall score. Lines present expected uncertainty per connection expressed in available bits of information about its default. *Red* are banks from domains 3–7; white banks from 2–7; yellow includes all banks

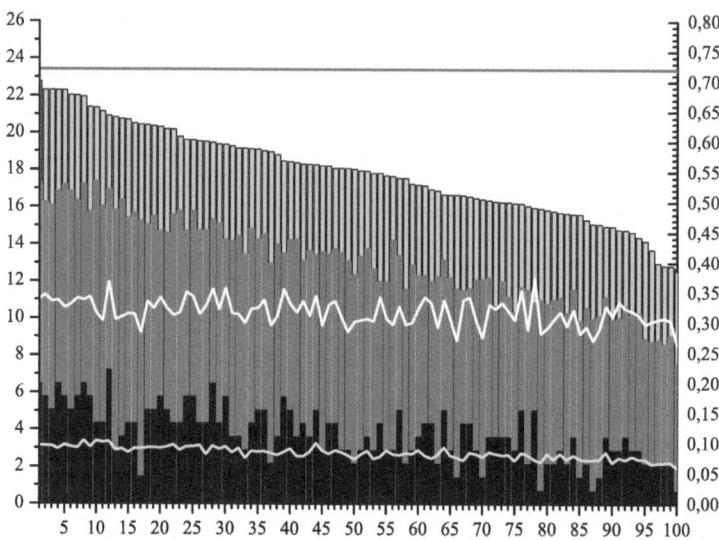

Fig. 6 A structure of stability index. Ordered descending per overall score. Scaling and topologies are the same as in Fig. 7, indexation of *bottom axis* presents ranks of banking networks, respectively

Prime result of this paper is a distribution of information about the default of any bank from banking system as measured by BI_r; see Eq. (8). According to the BI_r estimator, banking system 90 has potentially the most fragile connections (BI_{90} = 0.01616) and system 52 the least (BI_{58} = 0.00229); see Figs. 7–9 for more results).

4 Concluding Remarks

We have constructed a BI estimator for detecting a default potential of each unit of interbank liability (on average) within any banking system. The estimator is based on a composite influence estimator and a stability index and it depends on the quality of information about risk profiles of banks and information about their mutual exposures. The composite influence estimator is a measure for detecting the affinity of banking system to loosing its ability for providing liquidity (i.e. it is a combination of liquidity loss potential and α-criticality), while stability index measures a level of information that is available within the banking system about the default of any of its banks. The latter is based on Shannon's information entropy and can also serve as a tool for detecting the stability domain of banks and banking systems; we showed an example of its use in this respect and constructed seven fuzzy stability domains.

We tested the mechanism on a credit contagion model of 40 real banks connected in 100 random banking systems. Results are robust and the method can be easily employed to any banking system. Neither econometric nor statistical tests have

Banks and Their Contagion Potential: How Stable Is Banking System? 173

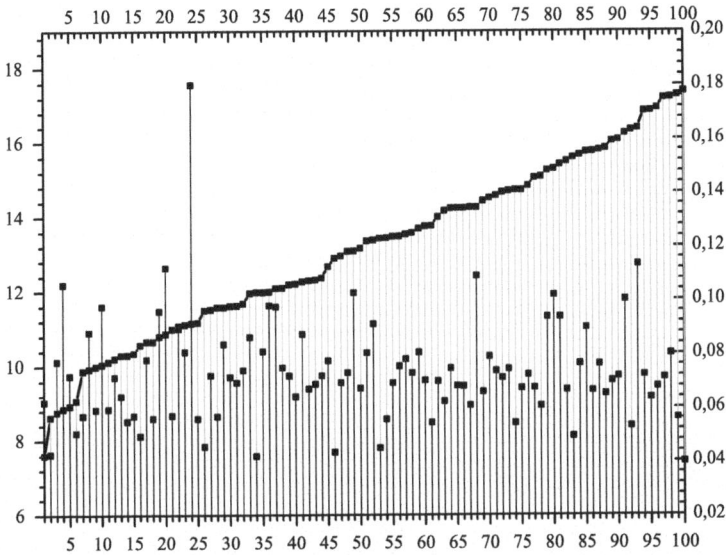

Fig. 7 Stability index (ascending order; *left axis*) and composite influence estimator (matched to stability index; *right axis*) of banking systems, given $p_i > 0.025$

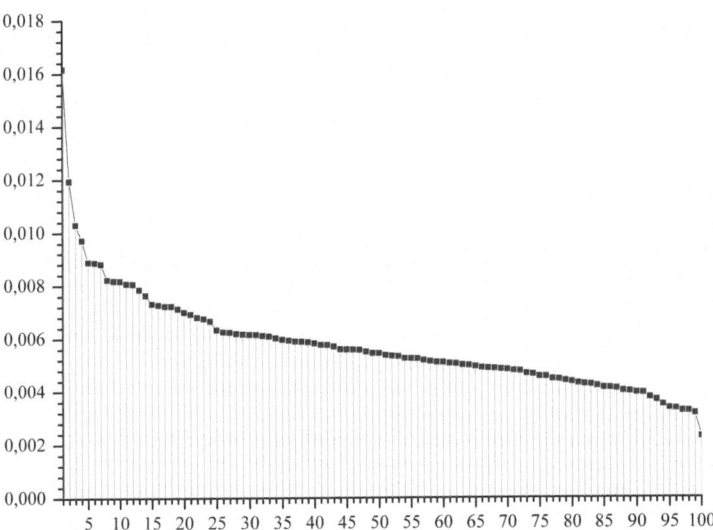

Fig. 8 Estimated contagion influence per one bit of information within the banking system (descending), given $p_i > 0.025$

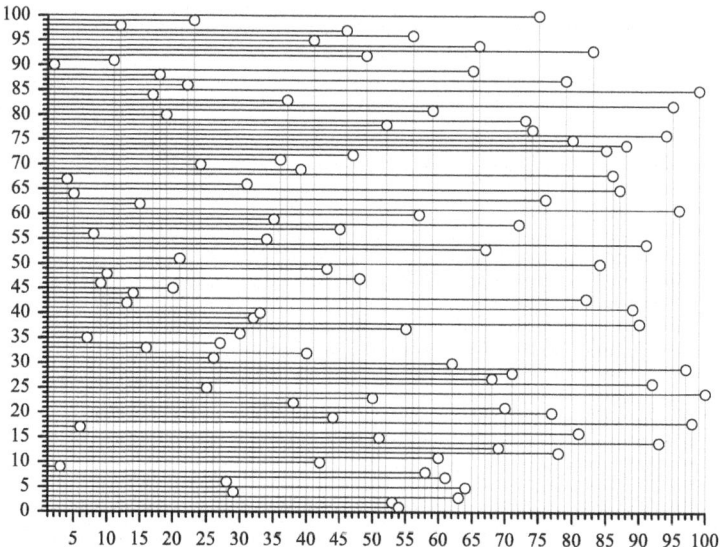

Fig. 9 Ranks of banking systems per their *BI* estimate (in descending order). *Left axis* shows identities of respective banking systems, *bottom axis* shows their ranks, accordingly

been performed on the final results. However, there is a huge potential to devise data-based statistical methods for studying the structural characteristics of banking system and its stability as driven by the network approach.

References

1. Allen F, Babus A (2008) Networks in finance. Available via SSRN. http://papers.ssrn.com/sol3/papers.cfm?abstract_id=1094883. Cited 30 Jan 2013.
2. Allen F, Gale D (2000) Comparing financial systems. MIT, Cambridge
3. Allen F, Babus A, Carletti E (2009) Financial crises: theory and evidence. Annu Rev Financ Econ 1:97–116
4. Boss M, Elsinger H, Summer M, Thurner S (2004) Network topology of the interbank market. Quant Financ 4(6):677–684
5. Cifuentes R, Ferrucci G, Shin H (2005) Liquidity risk and contagion. J Eur Econ Assoc 3:556–566
6. Dasgupta A (2004) Financial contagion through capital connections: a model of the origin and spread of bank panics. J Eur Econ Assoc 6:1049–1084
7. De Vries C (2005) The simple economics of bank fragility. J Bank Financ 29: 803–825
8. Eisenberg L, Noe T (2001) Systemic risk in financial systems. Manag Sci 47:236–249
9. Freixas X, Parigi B, Rochet J (2000) Systemic risk, interbank relations and liquidity provision by the central bank. J Money Credit Bank 32: 611–38
10. Gai P, Kapadia S (2010) Contagion in financial networks. Proc R Soc 466(2120):2401–2423
11. Haldane A, May R (2011) Systemic risk in banking ecosystems. Nature 469(7330):351–355

12. Iori G, De Masi G, Precup OV, Gabbi G, Caldarelli G (2008) A network analysis of the italian overnight money market. J Econ Dyn Control 32(1):259–278
13. Jorion P, Zhang G (2009) Credit contagion from counterparty risk. J Financ 64: 2053–2087
14. Leitner Y (2005) Financial networks: contagion, commitment, and private sector bailouts. J Financ 60:2925–2953
15. Schweitzer F, Fagiolo G, Sornette D, Vega-Redondo F, Vespignani A, White D (2009) Economic networks: the new challenges. Science 325:422–425
16. Steinbacher M, Steinbacher M, Steinbacher M (2012) Credit contagion in financial markets: a network-based approach. Available via SSRN. http://papers.ssrn.com/sol3/papers.cfm?abstract_id=2068716. Cited 30 Jan 2013
17. Upper C, Worms A (2004) Estimating bilateral exposures in the German interbank market: is there a danger of contagion? Eur Econ Rev 4:827–849

Part V
Organizations

Phrasing and Timing Information Dissemination in Organizations: Results of an Agent-Based Simulation

Doris A. Behrens, Silvia Berlinger, and Friederike Wall

Abstract This paper analyzes how managers suffering from decision-making biases in interrelated decision processes affect the performance of an overall business organization. To perform the analysis, we utilize an NK-type agent-based simulation model, in which decision-making is represented by adaptive walks on performance landscapes. We find that organizational performance holds up well, if the decision problem breaks into disjointed sub-problems. If decisions are, however, highly cross-related between departments, the overall organization's performance degrades, while both negatively phrasing information and relying more heavily on recently derived information account for an improvement. The effect of positively phrasing information that is relevant for decision-making works towards the same direction, but much more reluctantly. These results cautiously raise doubt about the claim that decision-making should *always* be as rational as possible.

D.A. Behrens (✉)
Department of Controlling and Strategic Management, Alpen-Adria Universität Klagenfurt, Universitätsstr. 65–67, 9020 Klagenfurt am Wörthersee, Austria
Department of Mathematical Methods in Economics, Research Unit for Operations Research and Control Systems, Vienna University of Technology, Argentinierstr. 8, 1040 Vienna, Austria
e-mail: doris.behrens@aau.at

S. Berlinger
Ilogs AG, Dammstr. 19, 6300 Zug, Switzerland
e-mail: silvia.berlinger@gmail.com

F. Wall
Department of Controlling and Strategic Management, Alpen-Adria Universität Klagenfurt, Universitätsstr. 65–67, 9020 Klagenfurt am Wörthersee, Austria
e-mail: friederike.wall@uni-klu.ac.at

1 Introduction

Descriptive decision theory shows that human behavior is influenced by cognitive biases that correspond to systematic deviations in judgments from what is 'rational' [24]. The recency effect accounts, for example, for the fact that individuals are more likely to remember information received only recently than information received some time ago (see, e.g., [3]). Thus, the timing of providing information to others matters. Another issue that matters is the way information is presented, how it is 'framed', and the associated effect is observed no matter whether information is rendered in a positive or in a negative way [25, 26]. While it is well-known that such biases trigger erroneous individual decisions, only little is known about how the presence of such biases, especially in combination with each other, impacts the performance of an overall organization consisting of several interdependent human decision makers. A notable exception comes from Berlinger and Wall [1] who employ an agent-based simulation for investigating decision-making processes affected by some prominent systematic errors as embodied in, e.g., the anchoring effect and the status quo bias.

Agent-based simulations constitute a powerful tool to analyze human decision-making behavior [2, 15], because this research method permits to investigate what would easily reach intractable dimensions in the case of formal modeling or empirical research (see [5] and [21], respectively). Therefore, relying on agent-based systems paves the way for gaining insight into decision-making structures within multidivisional organizations, different modes of coordination, the repercussions of altering levels of intra-organizational complexity, i.e., variations in interdepartmental relatedness, and into the effectiveness of diverse systems of providing incentives to departmental managers. Hence, agent-based systems allow for analyzing possible ramifications of combinations of recency and framing effects on the performance of a multidivisional organization with decentral decision-making. This, in particular, is why we utilize the computational model introduced by Berlinger and Wall [1], which is built upon Kauffman's NK model [7–10], and augment the model by simple 'local' rules describing systematic deteriorations of measuring individual performance provoked by the appearance of cognitive biases.[1]

2 The Agent-Based Simulation Model

To complement Berlinger and Wall [1]'s research by studying the repercussions of combinations of biases hitherto unaddressed, we set up a model that conceptualizes an organization as a system of interdependent choices [16].

---

[1] Note that the NK model has a substantial record of being successfully adopted for analyzing multidivisional organizations (see, e.g., [4, 17–20, 27]).

Let the departments of a multidivisional organization be endowded with full decision-making authority, none of them directly cooperating with any of the other departments. At each time step t ($t = 1, \ldots, T$), all m departments of the organization of concern have to come up with a total of N decisions, $d_{1t}, d_{2t}, \ldots, d_{Nt}$, where these decisions are modeled to be binary choices, i.e., $d_{it} \in \{0, 1\}$ ($i = 1, \ldots, N$). Then, an N-dimensional row vector $\mathbf{d}_t = [d_{1t} \ldots d_{Nt}]$ embodies a configuration of choices made at time t, where, due to the binary nature of the decision problem, 2^N different configurations are possible. The initial configuration of the decision problem, $\mathbf{d}_0 = [d_{10} \ldots d_{N0}]$, is generated randomly.

Plausibly, the choices made on the departmental level might easily affect the entire organization: at each time step t, each decision, d_{it}, has a share in the entire organization's performance, denoted by C_{it} (with $0 \leq C_{it} \leq 1$). The parameter K (with $K \in \{0, 1, \ldots, N-1\}$) accounts for the degree of intra-organizational interdependence and measures the extent to which decisions are related in their effects—even if departments neither deliberately compete nor cooperate.[2] If K were, for example, equal to zero, the performance landscape would have a single peak,[3] and each decision d_{it}'s contribution to organizational performance could be perceived as a function of d_{it} only, i.e., $C_{it} = f_{it}(d_{it})$. If K were equal to $N-1$, then the performance landscape would be maximally rugged and each decision would additionally affect the 'fitness value' of all other decisions, and d_{it}'s contribution to organizational performance would need to be defined as $C_{it} = f_{it}(d_{1t}, \ldots, d_{it}, \ldots, d_{Nt}) = f_{it}(\mathbf{d}_t)$. In other words, each decision, d_{it}, provides a contribution to organizational performance that additionally depends on K other decisions d_{nt} (with $n \neq i$), and we arrive at

$$C_{it} = f_{it}(d_{it}; K \text{ other } d_{nt}\text{'s}). \tag{1}$$

To account for these interdependencies between departments and the occurrence of spillover effects caused by the departments' decisions, for each performance contribution C_{it}, the payoff function $f_{it}(.)$ randomly draws a value from the uniform distribution $\mathcal{U}[0, 1]$. Whenever one of the interrelated decisions changes over time, the value for C_{it} ($0 \leq C_{it} \leq 1$) is redrawn. Assuming that $C_{1t}, C_{2t}, \ldots, C_{(N-1)t}$, and C_{Nt} equally contribute to overall organizational performance at time t, our performance measure corresponds to the normalized sum of its contributions, i.e.,

$$\mathcal{P}(\mathbf{d}_t) = \frac{1}{N} \cdot \sum_{i=1}^{N} C_{it}. \tag{2}$$

---

[2] For a thorough discussion of comprehensively mapping interaction consult [18]. Some additionally enlightning comments may be also found below Eq. 5.
[3] The model contains a fitness landscape representing the objective, the overall organizational performance, in which agents seek to incrementally improve their payoffs. This behavior corresponds to moving step by step from 'fitness valleys' to 'fitness peaks'.

Let then the departments have control only over a subset of the N decisions to be made, and let these subsets be disjointed and of equal size ($= N/m$). I.e., at each time step t, each department j ($j = 1, \ldots, m$) is held accountable for the sub-problem epitomized by deciding about what is contained in the N/m-dimensional row vector $\mathbf{d}_t^j = [d_{(l+1)t} \ldots d_{(l+N/m)t}]$ (with $l = (j-1) \cdot N/m$). Then, the stacked row vector

$$\mathbf{d}_t = [\mathbf{d}_t^1 \ldots \mathbf{d}_t^{j-1} \mathbf{d}_t^j \mathbf{d}_t^{j+1} \ldots \mathbf{d}_t^m] \tag{3}$$

of dimension N represents the configuration of choices made at time t, where each department j's actions are confined to its own sub-problem. When deciding about what to do, each department j is modeled to presume that all other departments ($k \neq j$) always stay with their previously optimal decisions, i.e., $\mathbf{d}_t^k = \mathbf{d}_{t-1}^{k*}$ (with $k = 1, \ldots, m; k \neq j$). Without any further coordination activity, each department j would, then, aim at incrementally improving its own performance, given by

$$\mathcal{P}^j(\mathbf{d}_t) = \frac{1}{N} \cdot \sum_{i=l+1}^{l+N/m} C_{it}, \quad l = (j-1) \cdot N/m, \tag{4}$$

by choosing the partial decision configuration $\mathbf{d}_t^{j*}$ that (in combination with the decision configurations $\mathbf{d}_t^k$, $k \neq j$, anticipated for all other departments) promises the highest individual utility; and j would do so without taking into account what is desired from an overall perspective. To align, however, selfish interests to organizational objectives,[4] the organization's central office—to some degree $\gamma^j \in [0,1]$—also rewards department j for the achievements of the other departments, $\mathcal{P}^k(.)$ (with $k = 1, \ldots, m; k \neq j$). Department j's value base for measuring the performance of a particular decision configuration $\mathbf{d}_t$ (and for getting compensated) is then defined by

$$\mathcal{B}^j(\mathbf{d}_t) = \mathcal{P}^j(\mathbf{d}_t) + \gamma^j \cdot \sum_{\substack{k=1 \\ k \neq j}}^{m} \mathcal{P}^k(\mathbf{d}_t). \tag{5}$$

Note that for $\gamma^j = 1$, individual and organizational objectives coincide, i.e., $\mathcal{B}^j(.) = \mathcal{P}^j(.) + \sum_{k(\neq j)=1}^m \mathcal{P}^k(.) = \mathcal{P}(.)$.

At each time step t, department j chooses the best out of a set of three partial configurations, which consists of (i) the *status quo*, i.e., $\mathbf{d}_t^j = \mathbf{d}_{t-1}^{j*}$, (ii) the *status quo* with an arbitrary digit inverted, and (iii) the *status quo* with two arbitrary digits inverted. If one of the newly discovered configurations (in combination with $\mathbf{d}_t^k = \mathbf{d}_{t-1}^{k*}$, $k \neq j$) improves j's performance, then this newly discovered

---

[4]This incentive structure corresponds to the one used, for example, in the model of Siggelkov and Rivkin [20] and the model of Berlinger and Wall [1].

Table 1 Self-contained structure for $N = 10$ and $m = 2$

	d_{1t}	d_{2t}	d_{3t}	d_{4t}	d_{5t}	d_{6t}	d_{7t}	d_{8t}	d_{9t}	d_{10t}
C_{1t}	×	×	×	×	×	–	–	–	–	–
C_{2t}	×	×	×	×	×	–	–	–	–	–
C_{3t}	×	×	×	×	×	–	–	–	–	–
C_{4t}	×	×	×	×	×	–	–	–	–	–
C_{5t}	×	×	×	×	×	–	–	–	–	–
C_{6t}	–	–	–	–	–	×	×	×	×	×
C_{7t}	–	–	–	–	–	×	×	×	×	×
C_{8t}	–	–	–	–	–	×	×	×	×	×
C_{9t}	–	–	–	–	–	×	×	×	×	×
C_{10t}	–	–	–	–	–	×	×	×	×	×

configuration becomes the next period's *status quo*, otherwise department j maintains its current *status quo*, i.e., $\mathbf{d}_t^{j*} = \mathbf{d}_t^j$ (cf. [14]).[5]

We know already that, due to several sorts of cross-departmental relationships, department j's choices may affect the contributions of other departments' decisions, and vice versa. Thus, whether or not department j's manager exclusively controls the department's performance depends on the degree of intra-organizational interdependence, K—and so does the manager's performance-based compensation (cf. Eq. 5).

In our model, we analyze two almost antagonistic types of intra-organizational relatedness (cf. [12, 13]). (i) A *self-contained structure* describes an organization, where departments are 'autarkic'. Then, department j' set of decisions, $d_{((j-1) \cdot N/m+1)t}, \ldots, d_{(j \cdot N/m)t}$, affects the set of performance contributions associated with its own sub-problem, i.e., $C_{((j-1) \cdot N/m+1)t}, \ldots, C_{(j \cdot N/m)t}$, to the highest possible extent, but not a single contribution of another department. In this case neither positive nor negative externalities are observed to occur between departments, the total decision problem breaks into m disjointed sub-problems, and $K = N/m - 1$. Table 1 illustrates this self-contained organizational structure for $N = 10$ and $m = 2$. (ii) In a *fully interrelated organizational structure* all decisions affect the performance contributions of all other decisions, i.e., $C_{it} = f_{it}(\mathbf{d}_t)$ ($i = 1, \ldots, N$) and we reach the highest possible degree of interrelatedness, i.e., $K = N - 1$.

This brings us to the very heart of our research topic, which is the analysis of biases that come into operation whenever departments choose the currently 'best-performing' decision configurations: Biased decision-making is modeled by assuming that each department j does not decide to choose a configuration that is actually best-performing but the configuration that is perceived as such. *Negative* or *loss framing*, for example, is modeled as altering the actual individual performance

---

[5]Note that the central office does not intervene in decentralized decision-making. It exercises influence only via providing incentives (by controlling γ^j). Moreover, the central office is obliged to note both departments' configurations ranked first, and, at the end of period t, to observe the performance of the combined configuration $\mathbf{d}_t^* = [\mathbf{d}_t^{1*} \ldots \mathbf{d}_t^{N*}]$.

measure, $\mathcal{B}^j(.)$ (cf. Eq. 5), in a way that a configuration $\mathbf{d}_t$ is misperceived such that it appears less attractive than it actually is, i.e.,

$$\mathcal{B}_L^j(\mathbf{d}_t) = \mathcal{B}^j(\mathbf{d}_t) \cdot (1 + \beta_L^j \cdot (\mathcal{B}^j(\mathbf{d}_t) - \mathcal{B}^{max})), \qquad (6)$$

where the term $(\mathcal{B}^j(\mathbf{d}_t) - \mathcal{B}^{max})$ in Eq. 6 is negative for all configurations that differ from the maximizing one. By phrasing information in such a way, the gap between what we have and what we could have is emphasized. The parameter $\beta_L^j > 0$ reflects department j's 'intensity' of responding to loss framing by weighting the influence of the gap on the perception of $\mathbf{d}_t$'s performance. *Positive* or *win framing* is modeled analogously and accounts for the fact that, by phrasing information, a configuration is misperceived such that it appears more attractive than it actually is, i.e.,

$$\mathcal{B}_W^j(\mathbf{d}_t) = \mathcal{B}^j(\mathbf{d}_t) \cdot (1 + \beta_W^j \cdot (\mathcal{B}^j(\mathbf{d}_t) - \mathcal{B}^{min})), \qquad (7)$$

where the parameter $\beta_W^j > 0$ accounts for department j's responsiveness to being informed about 'what can be gained compared to the minimum performance'.

If a *recency bias* is observed, the timing of providing information to the departments matters. I.e., the sequence in which decision configurations are discovered by the departments is relevant for decision-making, where in Eq. 8 a parameter, $\alpha^j > 0$, measures how strongly the timing of detecting configurations affects department j's perception of configuration $\mathbf{d}_t$'s performance. Recall that, in each time period, each department j discovers two alternative configurations of the partial decision vector. We assume that j gets to know all three configurations in random order, and, therefore, randomly assigns a label, $X_t(\mathbf{d}_t) \in \{1, 2, 3\}$, to each configuration, $\mathbf{d}_t$, which, in turn, shapes its perceived attractiveness as epitomized in

$$\mathcal{B}_R^j(\mathbf{d}_t) = \mathcal{B}^j(\mathbf{d}_t) \cdot (1 + \alpha^j \cdot (X_t(\mathbf{d}_t) - 3)). \qquad (8)$$

3 Simulation Setup

Let, without loss of generality, the multidivisional organization modeled here consist of a central office and $m = 2$ departments, where each of these has to come up with five decisions. Thus, $N = 10$.

For each simulation run, our artificial organization is observed for $T = 300$ time periods using 1,000 landscapes with five adaptive walks on each. The effect of cross-departmental interactions (as intensively discussed in [18]) is addressed by modeling decisions as being both solely related within departments ($K = 4$) and fully interrelated ($K = 9$). The former case is visualized by Table 1, where we postulate that an ×-entry indicates that contribution C_{it} ($i = 1, \ldots, 10$) is affected by decision d_{nt} ($n = 1, \ldots, 10$). The latter case would correspond to a table, where all boxes were filled with ×-entries.

We assume individual and organizational performance measures to either fully coincide (corresponding to a provision of firmwide incentives; $\gamma^j = 1$) or fully diverge (corresponding to departmental incentives; $\gamma^j = 0$), and all parameters quantifying the responsiveness to biases to be identical for both departments. Positive framing is modeled to occur with a likelihood of 70%, while negative framing is assumed to be evident in only 30% of all cases. We vary both the parameters β_L^j and β_W^j in the interval $I = [0, 0.2]$ in steps of 0.05. The intensity at which more recently received information is perceived to perform better than it actually does ranges from $\alpha^j = 0$ to 0.2 and is considered in steps of 0.05.

Note that the confidence intervals for the simulation results described in the forthcoming section vary between 0.006 and 0.007 at a confidence level of 99.9%.

4 Results and Interpretation

We derive all outcomes discussed in this section from inspecting $\mathcal{P}(\mathbf{d}_T^*)$, i.e., from the measure of organizational performance evaluated at the terminal period $T = 300$ (averaged over the simulation runs; not over time), where we do so in order to avoid an overvaluation of the transient part of the simulation. Recalling the measure's definition, we know that $\mathcal{P}(\mathbf{d}_t^*)$ is set up as the weighted sum of the performance contributions associated with all, to some degree interrelated, decisions, $\mathbf{d}_t^*$ ($t = 1, \ldots, T$). Thus, $0 \leq \mathcal{P}(\mathbf{d}_t^*) \leq 1$ ($t = 1, \ldots, T$). Therefore, a value of $\mathcal{P}(\mathbf{d}_t^*)$ close to one is a direct consequence of high average contribution levels and may be referred to as a 'good organizational performance', while $\mathcal{P}(\mathbf{d}_t^*) \ll 1$ stands for small average performance contributions and indicates a 'weak overall performance'.

As shown by Tables 2–4, we find that the organization's performance, as embodied in $\mathcal{P}(\mathbf{d}_{300}^*)$, is significantly better if the decision problem can be divided into two disjointed sub-problems (cf. [7, 28]).[6] Moreover, we discover that in the case of $\mathcal{P}(\mathbf{d}_{300}^*) \cong 0.97$, in the model presented here, there is hardly room for system improvement (corresponding to the robustness with respect to systematic errors being extraordinarily high), while the contrary can be observed in the case of an inferior organizational performance measure (i.e., for $\mathcal{P}(\mathbf{d}_{300}^*) \leq 0.90$)—where the latter typically corresponds to a more strongly interrelated decision problem. In this case the occurrence of systematic errors in decision-making caused by negatively phrasing information and relying more heavily on the most recently

---

[6]This finding is based on the fact that a higher degree of interrelation causes performances landscapes to be more rugged, which, in turn, makes it more difficult to find the global peak. A result resembling ours with respect to interrelatedness is provided by Stark and Behrens [22,23], who show within the context of an evolutionary game being played on a ring network that a lower degree of information and interaction accessible on the individual level (and thereby 'reducing the size' of the problem to be solved by the individual), respectively, can improve the performance of the overall system.

Table 2 Effect of positive framing (β_W^j) and negative framing (β_L^j) on $\mathcal{P}(\mathbf{d}_{300}^*)$

		Departmental incentives ($\gamma^j = 0$)				Firmwide incentives ($\gamma^j = 1$)				
		β_L^j				β_L^j				
		0.00	0.05	0.10	0.15	0.20	0.00	0.05	0.10	0.15	0.20
Part A: *Self-contained structure* ($K=4$)										
β_W^j	0.00	0.969	0.968	0.968	0.971	0.970	0.968	0.969	0.967	0.970	0.970
	0.05	0.970	0.970	0.969	0.968	0.970	0.970	0.970	0.969	0.969	0.970
	0.10	0.970	0.968	0.969	0.970	0.970	0.969	0.970	0.970	0.970	0.970
	0.15	0.970	0.970	0.968	0.971	0.970	0.970	0.970	0.970	0.969	0.969
	0.20	0.969	0.970	0.968	0.970	0.970	0.969	0.971	0.970	0.970	0.969
Part B: *Fully interrelated structure* ($K=9$)										
β_W^j	0.00	0.881	0.886	0.890	0.892	0.895	0.894	0.896	0.892	0.898	0.897
	0.05	0.884	0.887	0.890	0.895	0.895	0.894	0.896	0.896	0.897	0.895
	0.10	0.884	0.890	0.892	0.894	0.897	0.895	0.896	0.896	0.896	0.900
	0.15	0.885	0.889	0.894	0.893	0.899	0.896	0.898	0.896	0.898	0.897
	0.20	0.887	0.891	0.892	0.894	0.899	0.894	0.898	0.899	0.897	0.900

Table 3 Effect of recency bias (α^j) and positive framing (β_W^j) on $\mathcal{P}(\mathbf{d}_{300}^*)$

		Departmental incentives ($\gamma^j = 0$)				Firmwide incentives ($\gamma^j = 1$)				
		β_W^j				β_W^j				
		0.00	0.05	0.10	0.15	0.20	0.00	0.05	0.10	0.15	0.20
Part A: *Self-contained structure* ($K=4$)										
α^j	0.00	0.970	0.968	0.969	0.969	0.968	0.969	0.969	0.971	0.970	0.971
	0.05	0.971	0.971	0.970	0.970	0.969	0.971	0.971	0.969	0.971	0.970
	0.10	0.971	0.973	0.972	0.971	0.971	0.972	0.971	0.972	0.971	0.973
	0.15	0.972	0.971	0.973	0.972	0.971	0.971	0.972	0.971	0.973	0.970
	0.20	0.970	0.971	0.971	0.972	0.971	0.972	0.972	0.971	0.972	0.972
Part B: *Fully interrelated structure* ($K=9$)										
α^j	0.00	0.883	0.884	0.883	0.886	0.884	0.895	0.893	0.898	0.895	0.893
	0.05	0.894	0.894	0.895	0.894	0.892	0.896	0.899	0.896	0.899	0.898
	0.10	0.903	0.903	0.904	0.903	0.904	0.905	0.905	0.904	0.904	0.904
	0.15	0.911	0.911	0.911	0.910	0.911	0.912	0.913	0.911	0.911	0.911
	0.20	0.919	0.920	0.920	0.921	0.921	0.919	0.923	0.921	0.920	0.919

received information appears to be truly advantageous for the overall organization. The effect of positively phrasing information points into the same direction, but the results are profoundly less pronounced.

According to Table 2, Part B (and, in particular, for $\gamma^j = 0$), affecting decisions by encoding information in terms of 'something to lose' relative to the maximum level of the performance measure (cf. Eq. 6) significantly alters organizational performance, while shaping decisions by being informed about 'what can be gained compared to the minimum performance' (cf. Eq. 7) has no significant impact on

Table 4 Effect of recency bias (α^j) and negative framing (β_L^j) on $\mathcal{P}(\mathbf{d}_{300}^*)$

	Departmental incentives ($\gamma^j = 0$)					Firmwide incentives ($\gamma^j = 1$)				
	β_L^j					β_L^j				
	0.00	0.05	0.10	0.15	0.20	0.00	0.05	0.10	0.15	0.20
Part A: *Self-contained structure* ($K = 4$)										
0.00	0.969	0.969	0.968	0.968	0.971	0.968	0.968	0.969	0.968	0.969
0.05	0.970	0.971	0.969	0.971	0.971	0.972	0.972	0.970	0.871	0.972
α^j 0.10	0.971	0.971	0.970	0.971	0.971	0.972	0.972	0.972	0.972	0.970
0.15	0.972	0.971	0.971	0.971	0.970	0.971	0.971	0.971	0.971	0.971
0.20	0.971	0.972	0.972	0.972	0.971	0.972	0.972	0.970	0.971	0.971
Part B: *Fully interrelated structure* ($K = 9$)										
0.00	0.882	0.886	0.892	0.893	0.896	0.895	0.894	0.895	0.896	0.897
0.05	0.892	0.894	0.896	0.898	0.900	0.898	0.897	0.898	0.896	0.899
α^j 0.10	0.902	0.903	0.903	0.906	0.907	0.903	0.904	0.903	0.904	0.904
0.15	0.912	0.913	0.913	0.915	0.911	0.914	0.911	0.910	0.912	0.913
0.20	0.918	0.919	0.919	0.920	0.920	0.922	0.923	0.920	0.919	0.920

$\mathcal{P}(\mathbf{d}_{300}^*)$.[7] Recalling that positive framing is modeled to occur with a likelihood of 70%, while negative framing is assumed to be evident in only 30% of all cases, this observation translates into the finding that loss framing is much more effective—and prospect theory [6] comes to mind, from which we know that human decision makers tend to respond more strongly to something they 'lose', than to something they do not 'win' (cf. [11]).

Similar to a combined occurrence of positive framing and prioritizing more recently obtained information (see Table 3), the joint prevalence of negative framing and a recency bias (see Table 4) also substantially impacts organizational performance in a positive way, where the former result is mainly driven by the recency effect. For a highly interrelated decision problem experiencing the occurrence of the biases under investigation (with $\alpha^j = \beta_L^j = \beta_W^j = 0.2$), the terminal level of the organizational performance measure, $\mathcal{P}(\mathbf{d}_{300}^*)$, increases by almost 3% compared to the 'unbiased' problem—where, as depicted by Fig. 1, organizational performance mostly grows monotonically with the intensity parameters. Again, the necessary prerequiste for an improvement is that there is something to be improved at all, i.e., the problem has to be too highly interrelated to find the global peak in the performance landscape without any difficulty.

Consulting Tables 2–4, Part A, shows that changes in the incentives offered by the central office and embodied in γ^j have only a negligible effect on $\mathcal{P}(\mathbf{d}_{300}^*)$, if the decision problem breaks into disjointed sub-problems. Once the problem becomes, however, more complex (cf. Tables 2–4, Part B), the effect of providing

---

[7] Reading Table 2, it is plausible to assume that the non-significant effect of positive framing would turn into a significantly positive effect once the corresponding parameter range was extended, i.e., for $\beta_W^j > 0.2$.

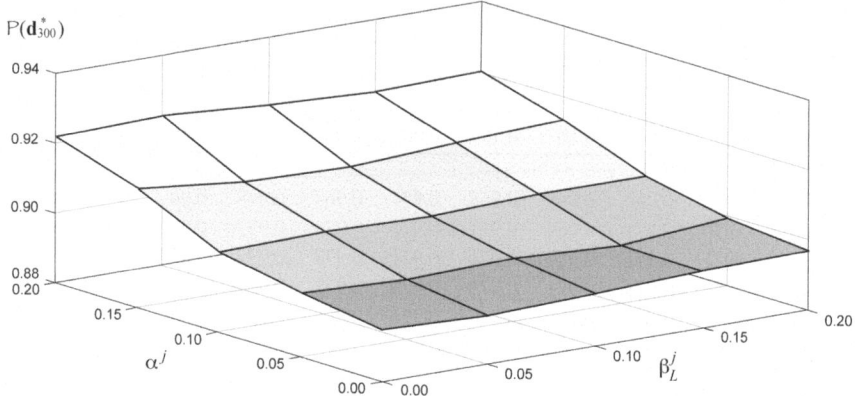

Fig. 1 Effect of recency bias (α^j) and negative framing (β_L^j) on $\mathcal{P}(\mathbf{d}_{300}^*)$ for $K = 9$

firmwide incentives to the departmental managers yields a significant improvement of organizational performance.

Altogether, our results suggest that biased decision-making allows for a more intensive exploration of the fitness/performance landscapes, i.e., it accomplishes a diversification of the search space (cf., e.g., [27]), and can, therefore, be superior to outcomes of fully rational decision-making. This is of particular importance in the case of intense interdepartmental relationships (affecting the interrelatedness of the decisions made).

5 Concluding Remarks

Utilizing an NK-type agent-based model it becomes obvious that the effects of decision-making biases vary with the complexity of the decision problem, and that the effects of the biases under investigation are most pronounced—especially if observed in combination with each other—if decisions are highly cross-related. Moreover, we find that organizational performance holds up well, if the decision problem breaks into disjointed (interrelated) sub-problems. If decisions are, however, highly cross-related between departments, the overall organization's performance degrades, while both negatively phrasing information and relying more heavily on recently derived information account for an improvement (especially if observed in combination with each other). The effect of positively phrasing information that is relevant for decision-making works into the same direction but much more reluctantly. These results cautiously raise the issue that with respect to decision-making non-rational behavior may well be superior to rational behavior.

Interesting perspectives for further research include, among others, an intensive investigation into the extent to which our results may be reversed when the intensities of the systematic biases become very large. In this spirit, simulations of non-systematic errors reveal that, once these errors become exhaustive, their beneficial effects disappear—and severe performance losses may even appear. Moreover, in the investigation presented here the application of 'firmwide incentives' is the sole mechanism of an artificial organization to coordinate departmental with organizational perspectives. Organizational design offers, however, an impressive range of other coordination mechanisms like, for example, lateral committees or vertical coordination. Hence, an interesting research question might be to find out to what extent far more intense coordination mechanisms could mitigate negative effects of (large scale) biases.

Further insights into the effectiveness of biased decision-making may also be derived from a closer investigation of the adaptive walks: One of the strengths of agent-based simulation is that it allows for examining the processes of adaptation, i.e., the processes of searching for higher levels of performance in our case. Thus, rather than confining the analyses to the (average) performance of the terminal period, it might be rather interesting to examine how decision-making biases affect processual aspects like, for example, the speed of performance enhancements or the percentage of false positive moves (which could induce costs).

Acknowledgements This work was carried out within the framework of the SOSIE project and was supported by Lakeside Labs GmbH. It was funded by the European Regional Development Fund (ERDF) and the Carinthian Economic Promotion Fund (KWF) under grant no. 20214/23793/35529.

References

1. Berlinger S, Wall F (2013) Effects of combined human decision-making biases on organizational performance. In: Amblard F, Giardini F (eds) Multi-agent-based simulation XII. Volume to appear of LNCS. Springer, Berlin/Heidelberg
2. Bonabeau E (2002) Agent-based modeling: methods and techniques for simulating human systems. Proc Natl Acad Sci USA 99(Suppl 3):7280–7287
3. Bredenkamp J, Wippich W (1977) Lern- und Gedächtnispsychologie I. Kohlhammer, Stuttgart
4. Chang MH, Harrington JE (2006) Agent-based models of organizations. In: Tesfatsion L, Judd KL (eds) Handbook of computational economics. Agent-based computational economics, vol 2. North Holland, Amsterdam, pp 1273–1337
5. Davis JP, Eisenhardt KM, Bingham CB (2007) Developing theory through simulation methods. Acad Manag Rev 32(2):480–499
6. Kahneman D, Tversky A (1979) Prospect theory: an analysis of decision under risk. Econometrica 47(2):263–292
7. Kauffman SA (1993) The origins of order. Self-organization and selection in evolution. Oxford University Press, New York
8. Kauffman SA (1995) At home in the universe. The search of laws for self-organization and complexity. Oxford University Press, New York
9. Kauffman SA, Levin S (1987) Towards a general theory of adaptive walks on rugged landscapes. J Theor Biol 128(1):11–45
10. Kauffman SA, Weinberger ED (1989) The NK model of rugged fitness landscapes and its application to maturation of the immune response. J Theor Biol 141:211–245

11. Leitner S, Behrens DA (2013) On the fault (in)tolerance of coordination mechanisms for distributed investment decisions: results of an agent-based simulation. Working paper, Alpen-Adria Universität Klagenfurt (in submission)
12. Leitner S, Wall F (2011) Effectivity of multi criteria decision-making in organisations: results of an agent-based simulation. In: Osinga S, Hofstede GJ, Verwaart T (eds) Emergent results of artificial economics. Volume 652 of LNEMS. Springer, Berlin/Heidelberg, pp 79–90
13. Leitner S, Wall F (2011) Unexpected positive effects of complexity on performance in multiple criteria setups. In: Hu B, Morasch K, Pickl S, Siegle M (eds) Operations research proceedings 2010 Munich. Springer, Berlin/Heidelberg, pp 577–582
14. Leitner S, Wall F (2013) Multi objective decision-making policies and coordination mechanisms in hierarchical organizations: results of an agent-based simulation. Working paper, Alpen-Adria Universität Klagenfurt (in submission)
15. Moss S (2001) Editorial introduction: Messy systems—the target for multi agent based simulation. In: Moss S, Davidsson P (eds) Multi-agent-based simulation. Volume 1979 of LNCS. Springer, Berlin/Heidelberg, pp 1–14
16. Porter ME (1996) What is strategy? Harv Bus Rev 74(6):61–78
17. Rivkin JW, Siggelkow N (2003) Balancing search and stability: interdependencies among elements of organizational design. Manag Sci 49(3):290–311
18. Rivkin JW, Siggelkow N (2007) Patterned interactions in complex systems: implications for exploration. Manag Sci 53(7):1068–1085
19. Siggelkow N, Levinthal DA (2003) Temporarily divide to conquer: centralized, decentralized, and reintegrated organizational approaches to exploration and adaptation. Organ Sci 14(6):650–669
20. Siggelkow N, Rivkin JW (2005) Speed and search: designing organizations for turbulence and complexity. Organ Sci 16(2):101–122
21. Sprinkle GB (2003) Perspectives on experimental research in managerial accounting. Account Organ Soc 28(2–3):287–318
22. Stark O, Behrens DA (2010) An evolutionary edge of knowing less (or: on the curse of global information). J Evol Econ 20(1):77–94
23. Stark O, Behrens DA (2011) In search of an evolutionary edge: trading with a few, more, or many. J Evol Econ 21(5):721–736
24. Tversky A, Kahneman D (1974) Judgment under uncertainty: heuristics and biases. Sci New Ser 185(4157):1124–1131
25. Tversky A, Kahneman D (1981) The framing of decisions and the psychology of choice. Sci New Ser 211(4481):453–458
26. Tversky A, Kahneman D (1986) Rational choice and the framing of decisions. J Bus 59(4):251–278
27. Wall F (2010) The (beneficial) role of informational imperfections in enhancing organisational performance. In: LiCalzi M, Milone L, Pellizzari P (eds) Progress in artificial economics. Computational and agent-based models. Volume 645 of LNEMS. Springer, Berlin/Heidelberg, pp 115–126
28. Weinberger E (1991) Local properties of Kauffman's N-K model: a tunably rugged energy landscape. Phys Rev A 44(10):6399–6413

On the Robustness of Coordination Mechanisms for Investment Decisions Involving 'Incompetent' Agents

Stephan Leitner and Doris A. Behrens

Abstract In this paper we transfer the concept of the competitive hurdle rate (CHR) mechanism introduced by Baldenius et al. (Account Rev 82(4):837–867, 2007) into an agent-based model, and test its robustness with respect to an occurrence of errors in forecasting. We find that our CHR born mechanism is most robust for highly diversified investment alternatives and a limited amount of those projects in need of scarce financial support. For misforecasting both the cash flow time series and the managers' individual efficiencies of operating investment projects, we find that this result reverses with an increasing extent of being wrong, so that a lower level of project heterogeneity appears to be more advantageous than a highly diversified investment landscape, i.e., if managers are really, really wrong about future economic development, the company fares better (or less worse, to be precise) if the investment alternatives are less dissimilar. This investigation allows to quantify the extent of error, when this comes about. Moreover, we provide policy advice for how an organization could design the framework of the CHR born mechanism so that forecasting errors, which inevitably occur, bring only minimal damage to the company.

S. Leitner (✉)
Faculty for Business and Economics, Department of Controlling and Strategic Management, Alpen-Adria Universität Klagenfurt, Klagenfurt, Austria
e-mail: stephan.leitner@aau.at

D.A. Behrens
Faculty for Business and Economics, Department of Controlling and Strategic Management, Alpen-Adria Universität Klagenfurt, Klagenfurt, Austria
Department of Mathematical Methods in Economics, Research Unit for Operations Research and Control Systems, Vienna University of Technology, Vienna, Austria
e-mail: doris.behrens@aau.at

1 Introduction and Research Question

It seems to be common knowledge that a corporate investment project should only be funded, if its internal rate of return exceeds some predetermined level, often called hurdle rate for obvious reasons. Nonetheless, research on how to appropriately pin down such a rate is rare. A remarkable exception comes from Baldenius, Dutta, and Reichelstein [2], who propose the competitive hurdle rate (CHR) mechanism for efficiently coordinating corporate investment decisions of multidivisional organizations with decentral decision-making taking account of multiple and often conflicting objectives. Their mechanism is designed in a market-like way (similar to a second price auction), and is particularly designed for optimally allocating scarce financial resources: Based on forecasts of essential project characteristics (initial cash outlay, cash flow series) and estimates of individual efficiencies of carrying out projects, an organization's central office announces individualized capital costs (which depend, among others, on the extent of intraorganizational competition for tight investment budgets, cf. [9]). Based on these costs of capital, department managers autonomously decide whether or not to implement the projects they presented to the central office in order to receive funding. The corresponding incentive system is set up such that exclusively efficient decisions, i.e., decisions that maximize the organization's shareholder value, are initiated.

Baldenius et al. [2] derive their CHR mechanism from an agency model, which incorporates some restrictive assumptions. Axtell [1] summarizes these assumptions as neoclassical sweetspot: full rationality, agent homogeneity, non-interactiveness, but also the availability of highly specific information [9]. Therefore, Eisenhardt [5] already dropped the subject that agency models are sometimes inappropriate for applications in the context of organizations, and Hendry [7] argues, moreover, that agents might not always be fully competent to achieve their objectives. I.e., also if agents undertake relatively simple tasks, following [7] they potentially make errors due to bounded rationality, limitations in foresight, and rational understanding [7, 13, 15]. Employing an agentization approach [6], we take account of these issues and transfer the concept behind the CHR mechanism introduced in Baldenius et al. [2] into an agent-based model.

Our agent-based model differs from Baldenius et al. [2]'s agency model along two essential lines. First, with respect to the agents' behavioral patterns, we follow Hendry [7], and model department managers such that they may be wrong when forecasting an investment alternative's main characteristics (initial cash outlay, cash flow time series), but also when they anticipate their own efficiency of operating a project. Second, since it has been widely discussed that the assumption of 'fully informed agents' might often be inappropriate to sufficiently epitomize actual human behavior in an organizational context (e.g., [5]), we additionally limit the agents' as well as the principal's information bases.

Coordinating Investment Decisions

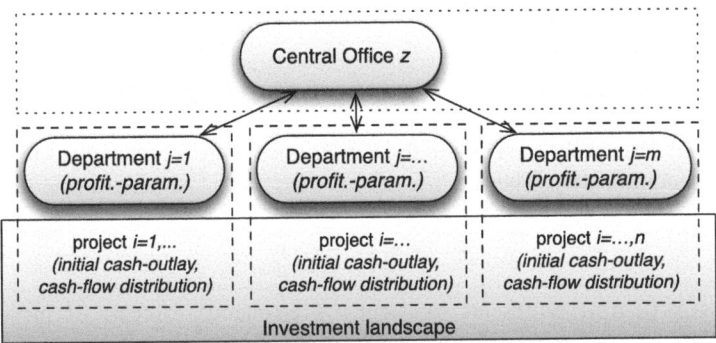

Fig. 1 Overview of the model with 'incompetent' agents

2 The Simulation Model

In our model, 'agents' correspond to *managers* or *department heads*, and the 'principal' to a *central office z* that is in charge of coordinative tasks and budget allocation. Each department's head j ($j = 1, \ldots, m$) is instructed to submit project proposals to the central office z, where a project's durableness is limited to T. Discovering a project is modeled as drawing an investment alternative i ($i = 1, \ldots, n$) from a randomly generated investment landscape (see, e.g., [11, 12]) that is characterized by an initial cash-outlay, κ_i (uniformly distributed in the interval $[\underline{\kappa}, \overline{\kappa}]$) and a cash-flow time series, $c_{it}/\sum_{\tau=1}^{T} c_{i\tau}$ for $t = 1, \ldots, T$ (c_{it} uniformly distributed in the unit interval). Moreover, the parameter η_i (uniformly distributed in $[\underline{\eta}, \overline{\eta}]$) defines i's profitability (e.g., i's return on investment), which is measured as a percentage of initial cash outlay, κ_i. Given the investment landscape's characteristics, for each investment alternative i the achievable cash-flow for time period t, thus, results in

$$\chi_{it} = \frac{\kappa_i \eta_i c_{it}}{\sum_{\tau=1}^{T} c_{i\tau}}. \tag{1}$$

In the model introduced by Baldenius et al. [2] agents are endowed with perfect foresight, and information regarding the alternative investment projects i is available for all managers j and the central office z. In the model presented here, we assume that decision makers are bounded rational and that each manager has information only on investment alternative that were discovered by himself or herself. The latter corresponds to disjointed information spaces as indicated by the dashed boxes in Fig. 1. Without loss of generality, we assume that a manager only submits the investment alternative that intra-departmentally scores highest. Then i denotes the sole project proposal and j stands for the manager proposing it, and we limit ourselves to the notion of i representing both the project submitted and the person who intends to carry it out. Then n and m also coincide, and we stay with the notion of n to indicate both the number of department heads and the number of serious project proposals.

When forecasting the main indicators of their projects, managers may be wrong [7]. We assume these non-systematical errors to be normally distributed with mean 0 and variance σ^2, and adjoin them multiplicatively to the undistorted values of the project indicators. Then,

$$\hat{\kappa}_i := \kappa_i \left(1 + \epsilon_{\kappa i}\right), \qquad (2)$$

with $\epsilon_\kappa \in \mathcal{N}\left(0, \sigma_\kappa^2\right)$, represents manager i's erroneous forecast of project i's initial cash-outlay. Correspondingly, manager i's error in observing the achievable cash-flow for period t is included in

$$\hat{\chi}_{it} := \chi_{it} \left(1 + \epsilon_{\chi it}\right), \qquad (3)$$

with $\epsilon_\chi \in \mathcal{N}\left(0, \sigma_\chi^2\right)$.

In the model presented here, we assume that managers are characterized by an efficiency of operating projects. Manager i's efficiency in carrying out a project, ρ_i (uniformly distributed in the interval $[\underline{\rho}, \overline{\rho}]$), is expressed as a fraction of the forecasted achievable cash-flows generated by project i. Since we consider departmental managers as being unable to come up with correct forecasts, the resulting error has to be taken into account, yielding

$$\hat{\rho}_i := \rho_i \left(1 + \epsilon_{\rho i}\right), \qquad (4)$$

with $\epsilon_\rho \in \mathcal{N}\left(0, \sigma_\rho^2\right)$.

The estimates $\hat{\kappa}_i$, $\hat{\chi}_{it}$, and $\hat{\rho}_i$ are next reported to the central office z (indicated by the solid arrows in Fig. 1), where z's information space is restricted to the managers' forecasts (indicated by the dotted box in Fig. 1).[1] Based on these, the central office calculates all projects' net present values (NPV),

$$\hat{\Lambda}_i = \sum_{t=1}^{T} \frac{\hat{\chi}_{it} \hat{\rho}_i}{(1+r)^t} - \hat{\kappa}_i, \quad i = 1, \ldots, n, \qquad (5)$$

where the parameter r represents the corporation's cost of capital. For each project i, z then computes a reference NPV, i.e., the highest NPV of all projects other than i, $\hat{\Lambda}_{-i}^* = max\{\hat{\Lambda}_1, \ldots, \hat{\Lambda}_{i-1}, \hat{\Lambda}_{i+1}, \ldots, \hat{\Lambda}_n\}$. Based on $\hat{\Lambda}_{-i}^*$, for every project i the central office individually calculates a capital charge rate, r_i^*, and reports it to

---

[1] Note that agency problems are excluded in the approach presented here. Thus, any deliberate misreporting by the departmental management can be ignored. Agents are incompetent but honest.

manager i. These *hurdle rates* are determined by exponential interpolation as they are implicitly defined by

$$\sum_{t=1}^{T} \frac{\hat{\chi}_{it}\hat{\rho}_i^*}{\left(1+r_i^*\right)^t} - \hat{\kappa}_i = 0, \tag{6}$$

with

$$\hat{\rho}_i^* := \hat{\rho}_i \cdot \frac{\hat{\Lambda}_{-i}^*}{\hat{\Lambda}_i} \tag{7}$$

being the level of efficiency for which manager i's project is at least as profitable as any of the other $n-1$ managers' projects (cf. also [2]). Whenever a departmental manager puts a proposed project into action, in every period t he or she is charged according to the relative benefit depreciation rule for the initial cash-outlay, κ_i, using the hurdle rate, r_i^*, as discount factor.[2] It then follows that residual income during time period t results in $\pi_{it} = \chi_{it} \cdot (\rho_i - \hat{\rho}_i^*)$ ([2], p. 850, cf. also [9]).

A manager is rewarded a fixed and a variable compensation component. The latter is determined by a function of residual income, $f(\pi_{it})$. The managers aim at maximizing the discounted stream of future variable compensation components. We assume that they discount at the principal's cost of capital, r. Based on the levels of the hurdle rates, managers decide whether or not to carry out their proposals, yielding the following decision-making rule (1 corresponds to 'invest', 0 corresponds to 'do not invest'),

$$I_i = \begin{cases} 1, & \text{if } \sum_{t=1}^{T} \frac{\hat{\chi}_{it}\hat{\rho}_i}{(1+r_i^*)^t} - \hat{\kappa}_i > 0 \\ 0, & \text{otherwise.} \end{cases} \tag{8}$$

Equation 8 implies that investment is attractive for only *one* project/manager.[3] Note that the computation of the hurdle rate is based on forecasted project indicators. Thus, the project finally realized may not necessarily be the one that is optimal seen from the central office's point of view.[4] This is caused by the managers' 'incompetence' and the central office's limited access to information that is free of errors.

---

[2] According to Rogerson [14], the relative benefit depreciation schedule at time t is calculated according to: $\frac{\hat{\chi}_{it}}{\sum_{\tau=1}^{T}(1+r_i^*)^{-\tau}\hat{\chi}_{i\tau}}$.

[3] Note that the optimization problem of manager i results in $\sum_{t=1}^{T} \frac{f(\pi_{it})}{(1+r)^t} \to max!$

[4] Considering the limitation of financial resources, the central office's optimization problem can be formalized as $\sum_{t=1}^{T} \frac{\chi_{it}\rho_i}{(1+r)^t} - \kappa_i - \sum_{t=1}^{T} \frac{f(\pi_{it})}{(1+r)^t} \to max!$ for $\sum_{i}^{n} I_i = 1$.

3 Simulation Experiments and Data Analysis

Christensen and Knudsen ([4], p. 84) propose that *'it is necessary to pursue research on the correlation between the economic context (project distribution), individual ability, and decision-making structure'*. We take note of this and, thus, parameterize the simulation model presented in the last section in the following way. First, we deal with project distribution by modeling organizations with diversified levels of project heterogeneity, $\mathcal{H}$, where we define $\mathcal{H} := \bar{\eta} - \underline{\eta}$. Thus, the higher $\mathcal{H}$ is, the more heterogenous is the investment landscape with respect to the projects' profitabilities, η_i, *ceteris paribus*. We fix the interval's lower boundary, $\underline{\eta}$, at 1.5 and vary the interval's upper boundary, $\bar{\eta}$, from 2.5 to 4.0 in steps of 0.5. Second, we investigate individual inabilities by varying the standard deviations of all possible forecasting errors from 0.05 to 0.3 in steps of 0.05. Third, we focus on the organizational rather than on decision-making structure. In particular, we investigate the degree of intra-organizational competition for scarce investment resources by varying the number of departments/managers/projects from $n = 2$ to $n = 6$ in steps of 2. All other parameters are kept constant, i.e., the assets' durableness, T, is fixed at 3; the central office's cost of capital, r, is set to 0.1; the initial cash outlay, κ_i, is drawn from $\mathcal{U}(100{,}000, 110{,}000)$; manager i's efficiency parameter, ρ_i, is drawn from $\mathcal{U}(0.80, 0.85)$. Considering all variations in the parameterization of the model, for each forecasting error $4 \times 6 \times 3 \times 3$ scenarios are investigated. For each scenario, we execute 80,000 simulation runs. Thus, the presented results are based on 1.728×10^7 simulations in total.

Let the project eventually put into action by the manager having proposed it to the central office be denoted by $\hat{\imath} := \{i \mid \sum_{t=1}^{T} \hat{\chi}_{it} \cdot \hat{\rho}_i \cdot (1 + r_i^*)^{-t} - \hat{\kappa}_i > 0\}$, and the optimal project be given by $i^* := \{i \mid max_i(\sum_{t=1}^{T} \chi_{it} \rho_i \cdot (1+r)^{-t} - \kappa_i)\}$.[5] In order to express the robustness of our CHR mechanism with respect to the managers' 'incompetence', we report two measures, i.e., the probability of operating suboptimal projects,

$$\tilde{P} := \mathbb{P}\left[\hat{\imath} \neq i^*\right], \tag{9}$$

and the foregone NPV to be expected,

$$\tilde{\Lambda} := \frac{1}{D} \sum_{d=1}^{D} \hat{\Lambda}_{\hat{\imath}}^d - \Lambda_{i^*}^d, \tag{10}$$

where the superscript d ($d = 1, \ldots, D$) indicates the simulation run. According to Eq. 5, $\Lambda_{i^*}^d = \sum_{t=1}^{T} \chi_{i^*t} \cdot \rho_{i^*} \cdot (1 + r)^{-t} - \kappa_{i^*}$, represents the shareholder value maximizing investment project for simulation run d.

---

[5]The selection of the finally realized project, $\hat{\imath}$, is based on forecasted values, while the shareholder value maximizing project, i^*, is determined on the basis of undistorted values.

Table 1 Forecasting error cash flow time series, $T = 3$

	Expected foregone NPV, $\tilde{\Lambda}$				Probability, $\mathbb{P}[i \neq i^*]$			
	Heterogeneity $\mathcal{H}$				Heterogeneity $\mathcal{H}$			
	1.00	1.50	2.00	2.50	1.00	1.50	2.00	2.50
Number of departments, $n = 2$								
0.05	−330.22	−282.89	−282.66	−272.58	0.0669	0.0504	0.0441	0.0386
0.10	−1,190.88	−1,055.56	−1,027.57	−1,010.13	0.1285	0.0982	0.0844	0.0737
0.15	−2,322.48	−2,172.86	−2,086.42	−2,134.90	0.1788	0.1422	0.1191	0.1073
0.20	−3,586.39	−3,533.76	−3,472.67	−3,498.66	0.2234	0.1814	0.1549	0.1380
0.25	−4,759.82	−4,852.44	−5,002.83	−5,183.01	0.2589	0.2125	0.1856	0.1690
0.30	−5,790.79	−6,280.71	−6,606.45	−6,874.11	0.2906	0.2424	0.2124	0.1942
Number of departments, $n = 4$								
0.05	−724.96	−700.37	−712.69	−707.28	0.0687	0.0554	0.0483	0.0442
0.10	−2,383.19	−2,357.29	−2,407.91	−2,476.02	0.1225	0.1016	0.0887	0.0811
0.15	−4,421.53	−4,414.96	−4,644.23	−4,780.80	0.1642	0.1374	0.1219	0.1121
0.20	−6,415.74	−6,655.18	−7,178.27	−7,344.25	0.1964	0.1659	0.1516	0.1376
0.25	−8,308.67	−8,810.05	−9,475.23	−10,170.13	0.2212	0.1902	0.1728	0.1608
0.30	−9,851.29	−11,013.01	−11,914.18	−12,934.46	0.2386	0.2102	0.1913	0.1802
Number of departments, $n = 6$								
0.05	−1,053.85	−1,004.92	−1,049.40	−1,106.03	0.0651	0.0522	0.0472	0.0434
0.10	−3,196.61	−3,252.72	−3,393.48	−3,567.55	0.1097	0.0929	0.0828	0.0767
0.15	−5,472.97	−5,841.79	−6,202.06	−6,678.08	0.1409	0.1219	0.1105	0.1032
0.20	−7,704.43	−8,310.50	−9,046.58	−9,850.64	0.1624	0.1431	0.1313	0.1235
0.25	−9,648.74	−10,735.66	−11,699.64	−12,814.45	0.1790	0.1594	0.1472	0.1396
0.30	−11,432.43	−12,884.80	−14,325.73	−15,672.88	0.1908	0.1720	0.1607	0.1516

(Std.dev. σ in leftmost column)

4 Results

From our simulation experiments, we gain some interesting insights about the drawbacks of distorted forecasts on the way the CHR mechanism is capable of successfully coordinating investment decisions. First of all, we find that errors in forecasting a manager's efficiency of operating a project (cf. Table 3) lead to the highest observed levels of distortion, followed by errors in forecasting the cash flow time series (cf. Table 1), and errors in forecasting the initial cash outlay (cf. Table 2). Most of the results presented in Tables 1–3 are significant in terms of confidence intervals (with $\alpha = 0.01$).

For errors in forecasting the cash flow time series and errors in forecasting the efficiency parameter, similar patterns are observed: For the case that errors are distributed with a low standard deviation, σ, increasing the level of heterogeneity, $\mathcal{H}$, leads to an increase in the level of the expected foregone NPV.[6] The opposite is

---

[6]Note that measured in absolute terms, the expected foregone NPV decreases.

Table 2 Forecasting error initial cash outlay, $T = 3$

		Expected foregone NPV, $\tilde{\Lambda}$				Probability, $\mathbb{P}[i \neq i^*]$			
		Heterogeneity $\mathcal{H}$				Heterogeneity $\mathcal{H}$			
		1.00	1.50	2.00	2.50	1.00	1.50	2.00	2.50
	Number of departments, $n = 2$								
Std.dev. σ	0.05	−276.46	−190.66	−153.29	−114.87	0.0624	0.0426	0.0337	0.0260
	0.10	−1,017.11	−725.95	−566.92	−452.02	0.1199	0.0836	0.0649	0.0506
	0.15	−2,018.40	−1,537.15	−1,202.33	−1,002.13	0.1659	0.1209	0.0925	0.0760
	0.20	−3,182.31	−2,553.49	−2,032.42	−1,711.97	0.2132	0.1551	0.1206	0.0986
	0.25	−4,319.18	−3,599.93	−2,990.41	−2,541.71	0.2474	0.1855	0.1442	0.1205
	0.30	−5,401.91	−4,763.03	−4,051.50	−3,447.66	0.2789	0.2123	0.1693	0.1401
	Number of departments, $n = 4$								
Std.dev. σ	0.05	−518.34	−368.80	−268.47	−229.95	0.0578	0.0407	0.0298	0.0252
	0.10	−1,890.86	−1,360.08	−1,066.37	−876.59	0.1085	0.0770	0.0599	0.0486
	0.15	−3,715.01	−2,838.75	−2,259.48	−1,871.08	0.1497	0.1099	0.0861	0.0704
	0.20	−5,752.25	−4,576.42	−3,798.56	−3,111.43	0.1846	0.1368	0.1107	0.0898
	0.25	−7,650.25	−6,611.13	−5,508.97	−4,732.70	0.2105	0.1633	0.1320	0.1100
	0.30	−9,649.57	−8,665.24	−7,557.49	−6,459.94	0.2332	0.1845	0.1527	0.1276
	Number of departments, $n = 6$								
Std.dev. σ	0.05	−283.94	−192.97	−145.91	−116.56	0.0545	0.0385	0.0297	0.0234
	0.10	−1,012.94	−721.36	−547.18	−467.05	0.0967	0.0708	0.0559	0.0459
	0.15	−2,021.23	−1,485.19	−1,195.72	−989.72	0.1284	0.0978	0.0784	0.0656
	0.20	−3,188.82	−2,522.01	−2,022.43	−1,668.80	0.1532	0.1194	0.0977	0.0828
	0.25	−4,260.87	−3,593.90	−2,998.51	−2,578.79	0.1711	0.1387	0.1149	0.0979
	0.30	−5,342.65	−4,754.71	−4,078.54	−3,453.13	0.1874	0.1531	0.1297	0.1116

observed for high standard deviations, where the absolute value of the expected foregone NPV, $|\tilde{\Lambda}|$, increases with increasing $\mathcal{H}$. Thus, a 'point of inflexion', denoted by $\mathcal{P}$, occurs. I.e., to minimize the odd effects of forecasting errors, for $\sigma > \mathcal{P}$ we find that lower $\mathcal{H}$-values should be preferred over higher $\mathcal{H}$-values. Moreover, our results indicate that the 'point of inflexion' shifts into the direction of larger σ-values, if the number of serious project proposals, n, increases.

Above findings are due to the fact that relatively small errors do *not* cause any reranking of the proposed projects. Relatively large errors lead, however, more frequently to project proposals being wrongly reranked with respect to their contributions to shareholder wealth. Increases in project heterogeneity, $\mathcal{H}$, imply that the 'differences' between the projects become larger. In contrast, relatively small σ-values do not affect the ranking if projects are 'very different', which is why the errors' undesired effects decrease with increases in $\mathcal{H}$. In contrast, relatively large σ-values 'overcome the differences' between projects, which leads to an increase of the (absolute value of) expected foregone NPV, $|\tilde{\Lambda}|$. In order to illustrate the matter, let us outline an example (cf. Table 1, $n = 2$): For a low standard deviation, i.e., $\sigma = 0.05$, $|\tilde{\Lambda}|$ computed for errors in forecasting the cash flow time series decreases from 330.22 to 272.58 as the degree of heterogeneity, $\mathcal{H}$, increases

Coordinating Investment Decisions

Table 3 Forecasting error departmental efficiency, $T = 3$

Foregone NPV, $\tilde{\Lambda}$					Probability, $\mathbb{P}[\mathrm{i} \neq i^*]$			
	Heterogeneity $\mathcal{H}$				Heterogeneity $\mathcal{H}$			
	1.00	1.50	2.00	2.50	1.00	1.50	2.00	2.50
Number of departments, $n = 2$								
0.05	−732.64	−633.90	−616.30	−605.71	0.1008	0.0765	0.0651	0.0579
0.10	−2,442.38	−2,232.00	−2,131.93	−2,161.88	0.1852	0.1446	0.1205	0.1089
0.15	−4,300.98	−4,400.32	−4,352.38	−4,428.46	0.2444	0.2010	0.1743	0.1552
0.20	−5,911.91	−6,335.79	−6,839.38	−7,127.53	0.2916	0.2436	0.2168	0.1989
0.25	−7,206.81	−8,344.47	−9,064.68	−9,727.46	0.3256	0.2818	0.2535	0.2330
0.30	−8,245.10	−9,858.64	−11,075.46	−12,156.77	0.3529	0.3110	0.2802	0.2613
Number of departments, $n = 4$								
0.05	−1,545.19	−1,464.90	−1,511.47	−1,524.81	0.1003	0.0800	0.0710	0.0635
0.10	−4,541.71	−4,627.56	−4,717.46	−4,975.51	0.1674	0.1402	0.1236	0.1136
0.15	−7,530.51	−7,953.67	−8,544.26	−8,932.54	0.2117	0.1798	0.1646	0.1497
0.20	−10,037.87	−11,264.49	−12,119.05	−13,180.33	0.2406	0.2138	0.1933	0.1829
0.25	−12,120.89	−14,052.76	−15,588.19	−17,039.86	0.2630	0.2344	0.2181	0.2043
0.30	−13,686.27	−16,398.51	−18,397.32	−20,474.77	0.2777	0.2521	0.2326	0.2236
Number of departments, $n = 6$								
0.05	−2,139.60	−2,133.04	−2,210.07	−2,305.38	0.0912	0.0761	0.0682	0.0620
0.10	−5,610.86	−6,014.24	−6,410.39	−6,785.39	0.1431	0.1243	0.1129	0.1048
0.15	−8,952.02	−9,804.65	−10,684.70	−11,781.61	0.1739	0.1538	0.1423	0.1352
0.20	−11,503.24	−13,271.66	−14,655.92	−15,992.54	0.1921	0.1749	0.1621	0.1545
0.25	−13,649.69	−15,987.27	−18,227.56	−20,167.55	0.2052	0.1882	0.1784	0.1694
0.30	−15,293.43	−18,398.87	−21,002.57	−23,606.16	0.2144	0.1990	0.1884	0.1816

from 1.0 to 2.5. A reversal of this pattern can be observed for $\sigma = 0.25$, where the absolute value of the expected foregone NPV begins to increase with increases in $\mathcal{H}$, i.e., for $\sigma = 0.25$ and $\mathcal{H} = 1.0$, $|\tilde{\Lambda}| = 4{,}759.82$. As $\mathcal{H}$ increases to 2.5, the absolute value of the expected foregone NPV increases to 5,183.01.

For the probability of operating a suboptimal project, we find that it decreases with increases in project diversity, $\mathcal{H}$. According to Leitner and Behrens [10], similar patterns for the foregone NPV as well as for $\tilde{P}$ can be observed for variations in assets' useful life, T. For errors in the forecasts of a manager's efficiency of operating a project as well as for misforecasts of the cash flow time series, this allows us to conclude that our implementation of the CHR mechanism is most robust for small n- and high $\mathcal{H}$-values. Also computing the expected values of the foregone NPV (i.e., $|\tilde{\Lambda}| \cdot \tilde{P}$) allows for this conclusions. The pattern which can be observed for errors in forecasting cash flow time series is sketched in Fig. 2. Apart from the magnitude of foregone NPVs, errors in forecasting the efficiency of operating projects lead to a very similar pattern as misforecasting cash flow time series does.

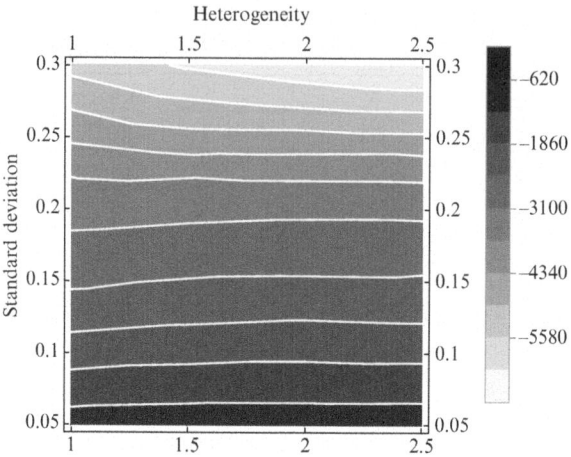

Fig. 2 Forecasting error in cash flow time series ($n = 2$)

For forecasting errors of the initial cash outlay, within the investigated ranges of $\mathcal{H}$ and σ (cf. Table 2), our results indicate that the absolute level of foregone NPV, $|\tilde{\Lambda}|$, decreases with increases in the project diversity, $\mathcal{H}$. Up to a σ-level of 0.3 we do not observe a 'point of inflexion', as is the case for the two other types of errors. What is, however, in line with the errors in forecasting a manager's efficiency and the cash flow series: The probability to operate a suboptimal project decreases with increases in $\mathcal{H}$. Moreover, our results suggest that the difference between the absolute values of the forgone NPV in the case of $\mathcal{H} = 2.5$ and $\mathcal{H} = 1.0$ increases with increasing σ. This indicates that potentially there exists a 'point of inflexion' for higher σ-values. Nonetheless, increasing the standard deviation would lead to unrealistically high errors.[7] The pattern which can be observed for misforecasting the initial cash outlay is delineated in Fig. 3.

To sum up our results: homogenous investment alternatives are less robust to forecasting errors than investment landscapes that are characterized by a higher level of project diversity. For errors in forecasting the initial cash outlay, this finding is independent from the extent of being wrong, σ, and the number of serious project proposals, n. For misforecasting a manager's efficiency parameter and a project's cash flow series, we observe a 'point of inflexion', which, with increases in n, shifts towards larger σ-values. For σ-values $< \mathcal{P}$, higher $\mathcal{H}$-values are to be preferred over lower levels of heterogeneity. For σ-values $> \mathcal{P}$ the opposite is true.

This allows for providing some important policy advice: in the short run, the level of project heterogeneity can be regarded as an element of choice, i.e., the corporation can, for example, toughen the minimum requirements with respect to

---

[7]Notice that an error which is ≥ 1 would render the initial cash outlay to a cash inflow, which is far away from reality.

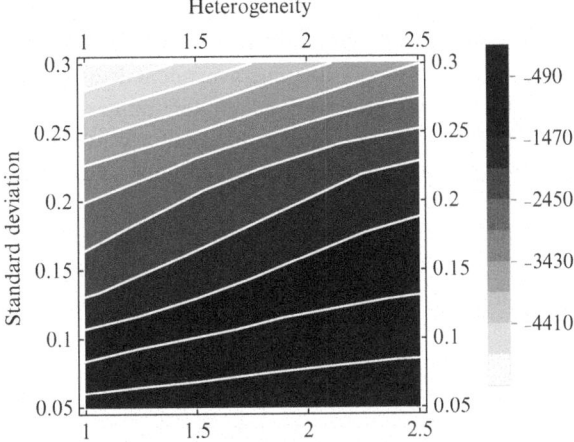

Fig. 3 Forecasting error in initial cash outlay ($n = 2$)

profitability that intended projects have to fulfill in order to be accepted as a proposal for funding—and, at the same time, limit project diversity. If, in addition, the central office has knowledge on the extent of 'departmental incapability', the contour plots given in Figs. 2 and 3 may constitute a solid basis, if one has to come up with the decision about when to narrow down the level of project diversity in order to minimize the effects of incorrect forecasts. However, it has to be kept in mind that, besides minimizing the errors' effects, limiting the level of heterogeneity also limits the corporation's future development potential (cf. also [9]).

5 Concluding Remarks

In this paper we describe how we transfer the concept of the competitive hurdle rate (CHR) mechanism, which is introduced by Baldenius, Dutta, and Reichelstein [2] to achieve efficient budget allocation, into an agent-based model, and test the robustness of our CHR born mechanism with respect to errors in forecasting. We find that the robustness of the mechanism critically depends on two parameters: the level of project diversity, and the number of project proposals. The CHR mechanism turns out to be most robust for a high level of heterogeneity, and a relatively low number of projects competing for scarce funding. For errors in forecasting the cash flow time series, and the managers' efficiencies in operating projects, we find that with increases in the extent of error, this finding reverses, so that a lower level of heterogeneity appears to be beneficial over a diverse investment landscape. For organizations, the level of heterogeneity is an important policy parameter that allows for dampening the negative effects of forecasting errors. We provide policy advice for how to design framework conditions, so that forecasting errors lead to a minimal magnitude of foregone net present value (NPV).

Our simulation model might be extended in the following ways: the investigation of more than one error at a time would give important insights into how to efficiently design framework conditions for coordinating investment decisions. Thus, the effect of combinations of errors is one potential avenue for future research (for research on the combination of errors cf., for example, [3, 8]). For some of our variables, we assume a uniform distribution. Future research might test whether the assumptions regarding the distributions affect the results. Moreover, communication among agents is assumed to be cut off. As some local interactions could positively affect the forecasting quality (cf., [16, 17]), one further promising avenue for future research would be to complement the model by intraorganizational communication within a network structure.

Acknowledgements A part of Doris A. Behrens' work was carried out within the framework of the SOSIE project and was supported by Lakeside Labs GmbH. It was funded by the European Regional Development Fund (ERDF) and the Carinthian Economic Promotion Fund (KWF) under grant no. 20214/23793/35529.

References

1. Axtell RL (2007) What economic agents do: how cognition and interaction lead to emergence and complexity. Rev Austrian Econ 20(2–3):105–122
2. Baldenius T, Dutta S, Reichelstein S (2007) Cost allocation for capital budgeting decisions. Account Rev 82(4):837–867
3. Behrens DA, Berlinger S, Wall F (2013) Phrasing and timing information dissemination in organizations: results of an agent-based simulation. In: Leitner S, Wall F (eds) Artificial economics and self-organization. LNEMS. Springer, Berlin/Heidelberg/New York, pp 179–190
4. Christensen M, Knudsen T (2007) Design of decision-making organizations. Manag Sci 56(1):71–89
5. Eisenhardt KM (1989) Agency theory: an assessment and review. Acad Manag Rev 14(1):57–74
6. Guerrero OA, Axtell RL (2011) Using agentization for exploring firm and labor dynamics. A methodological tool for theory exploration and validation. In: Osinga S, Hofstede GJ, Verwaart T (eds) Emergent results of artificial economics. Volume 652 of LNEMS. Springer, Heidelberg/Berlin/New York, pp 139–150
7. Hendry J (2002) The principal's other problems: honest incompetence and the specification of objectives. Acad Manag Rev 27(1):98–113
8. Leitner S (2012) A simulation analysis of interactions among intended biases in costing systems and their effects on the accuracy of decision-influencing information. Cent Eur J Oper Res. doi:10.1007/s10100-012-0275-2
9. Leitner S, Behrens DA (2013) Is it or is it not a good idea to derive coordination mechanisms from collusion-free agency models? Consequences for residual income measurement. Working paper, Alpen-Adria Universität Klagenfurt
10. Leitner S, Behrens DA (2013) On the fault in tolerance of coordination mechanisms for investment decisions: results of an agent-based simulation. Working paper, Alpen-Adria Universität Klagenfurt

11. Leitner S, Wall F (2011) Effectivity of multi criteria decision-making in organisations: results of an agent-based simulation. In: Osinga S, Hofstede GJ, Verwaart T (eds) Emergent results of artificial economics. Volume 652 of LNEMS. Springer Berlin Heidelberg, pp 79–90
12. Leitner S, Wall F (2011) Unexpected positive effects of complexity on performance in multiple criteria setups. In: Hu B, Morasch K, Pickl S, Siegle M (eds) Operations research proceedings 2010. Springer, Berlin/Heidelberg, pp 577–582
13. Martin R (1993) The new behaviorism: a critique of economics and organization. Hum Relat 46(9):1085–1101
14. Rogerson WP (1997) Intertemporal cost allocation and managerial investment incentives: a theory explaining the use of economic value added as a performance measure. J Pol Econ 105:770–795
15. Simon HA (1955) A behavioral model of rational choice. Q J Econ 69(1):99–118
16. Stark O, Behrens DA (2010) An evolutionary edge of knowing less (or: On the curse of global information). J Evol Econ 20(1):77–94
17. Stark O, Behrens DA (2011) In search of an evolutionary edge: trading with a few, more, or many. J Evol Econ 21(5):721–736

Pack Light on the Move: Exploitation and Exploration in a Dynamic Environment

Marco LiCalzi and Davide Marchiori

Abstract This paper revisits a recent study by Posen and Levinthal (Manag Sci 58:587–601, 2012) on the exploration/exploitation tradeoff for a multi-armed bandit problem, where the reward probabilities undergo random shocks. We show that their analysis suffers two shortcomings: it assumes that learning is based on stale evidence, and it overlooks the steady state. We let the learning rule endogenously discard stale evidence, and we perform the long run analyses. The comparative study demonstrates that some of their conclusions must be qualified.

1 Introduction

In many situations, an agent must simultaneously make decisions to maximize its rewards while learning the process that generates these rewards. This leads to a tradeoff between exploration versus exploitation. Exploratory actions gather information and attempt to discover profitable actions. Exploitative actions aim to maximize the current reward based on the present state of knowledge. When the agent diverts resources towards exploration, he sacrifices the current reward in exchange for the hope of higher future rewards.

The dilemma between exploration and exploitation is well-known in machine learning, where the agent is an algorithm; see f.i. Cesa-Bianchi and Lugosi [2]. Within this field, the simplest and most frequent example is the multi-armed bandit problem, extensively studied in statistics as well [1]. However, in the literature on organizational studies, the exploration/exploitation trade-off has come to be associated mostly with a seminal contribution by March [5], that introduced a peculiar model of his own.

M. LiCalzi (✉) · D. Marchiori
Department of Management, Università Ca' Foscari Venezia, Venice, Italy
e-mail: licalzi@unive.it; davide.marchiori@unive.it

The popularity of March [5], as witnessed by more than 10,000 citations on Google Scholar, has firmly placed the exploration/exploitation trade-off among the methodological toolbox of organizational studies, but the peculiarity of his modeling choice has shifted attention away from the multi-armed bandit problem as a modeling tool. This shortcoming was recently addressed by Posen and Levinthal [6], that explicitly discuss some similarities between the bandit problem and the March model.

Their paper inquires about the implications of the exploration/exploitation trade-off for organizational learning when the environment changes dynamically or, more precisely, when the process generating the rewards is not stationary. Using the bandit problem as a workhorse, they challenge the conventional view that an increasingly turbulent (i.e., non-stationary) environment should necessarily elicit more exploration.

We believe that Posen and Levinthal [6] make two very important contributions. First, they raise fundamental questions (as well as providing convincing answers) about the impact of turbulence in an environment for organizational learning. Second, they implicitly make a strong methodological case for a revival of the bandit problem as a modeling tool.

On the other hand, we argue that two (apparently minor) of their modeling choices are potentially misleading. The first one is the length of the horizon over which the study is carried out: this is too short to provide information about the steady state. The second one is that learning is based on the whole past evidence (including what turbulence has made obsolete): this makes it too slow to detect shocks, and hence ineffective.

This paper sets out to discuss and correct these flaws, revisiting their analysis over the short and the long run. We propose two (nested) learning models that endogenously recognize and shed away stale evidence, and compare their performance with the original model by Posen and Levinthal [6]. We check several of their conclusions, and show how a few of these need to be qualified. Paraphrasing the title of their paper, our major result demonstrates the importance of packing light (evidence) when chasing a moving target. Shedding away obsolete information is crucial to attain a superior performance as well as making learning resilient to shocks.

2 The Model

We summarize the model proposed in Posen and Levinthal [6]; then, we present the crucial tweaks we advocate. At each period t, an organization must choose among $N = 10$ alternatives. Each alternative $i = 1, \ldots, 10$ has two possible outcomes: $+1$ (success) or -1 (failure). These are generated as a (Bernoullian) random reward R_t^i in $\{-1, 1\}$, with probability p_t^i of success. Thus, the state of the environment in period t is summarized by the vector $P_t = [p_t^1, \ldots, p_t^{10}]$.

In the standard bandit problem, the environment is stationary and $P_t = P$ for all t. Posen and Levinthal [6]—from now, PL for brevity—relax this assumption and introduce environmental turbulence as follows. Each alternative i is given an initial probability p_0^i randomly drawn from a Beta distribution with $\alpha = \beta = 2$. This has a unimodal and symmetric density, with expected value $1/2$ and variance $1/20$. The turbulence in the environment follows from a probabilistic shock that may occur in each period with probability η. When $\eta = 0$, the environment is stationary; increasing η raises the level of turbulence. For $\eta > 0$, PL assume $\eta = 0.005 \times 2^k$ with k being an integer between 0 and 6. When a shock occurs, each of the payoff probabilities is independently reset with probability $1/2$ by an independent draw from the same Beta distribution.

At each period t, the organization holds a propensity q_t^i for each alternative that is formally similar (and proportional to) its subjective probability assessment that the i-th alternative yields success, and thus leads to a reward of 1. At time t, its propensities over the 10 available alternatives are summarized by the vector $Q_t = [q_t^1, \ldots, q_t^{10}]$. Propensities are updated using a simple rule, akin to similar treatments in reinforcement learning; see Duffy [3].

Let n_t^i be the number of successes and the total number of plays for the i-th alternative up to (and including) period t. PL define the propensities recursively by

$$q_{t+1}^i = \left(\frac{n_t^i}{n_t^i + 1}\right) q_t^i + \left(\frac{1}{n_t^i + 1}\right) \frac{R_t^i + 1}{2} \tag{1}$$

with the initial condition $q_0^i = 1/2$ for each i. As n_t^i increases, the weight associated to the most recent outcome declines.

This paper follows PL's assumption about propensities to facilitate comparison. However, we notice that Eq. (1), while certainly reasonable, is a reduced form that omits the specification of the relationship between q_t^i and the number of successes and failures experienced with the i-th alternative. A more explicit formulation might have been the following. Let s_t^i and n_t^i be respectively the number of successes and the total number of plays for the i-th alternative up to (and including) period t. Let us define the propensities by $q_{t+1}^i = (1 + s_t^i)/(2 + n_t^i)$, with the initial condition $s_0^i = n_0^i = 0$ to ensure $q_1^i = 1/2$ for each i. Then the updating rule for propensities would read

$$q_{t+1}^i = \left(\frac{n_t^i + 1}{n_t^i + 2}\right) q_t^i + \left(\frac{1}{n_t^i + 2}\right) \frac{R_t^i + 1}{2}$$

The choice behavior in each period depends on the distribution of propensities and on the intensity of the search strategy. More precisely, PL assume a version of the *softmax* algorithm; see f.i. Sutton and Barto [7]. In period t, the organization picks alternative i with probability

$$m_t^i = \frac{\exp\left(10 q_t^i / \tau\right)}{\sum_{j=1}^{10} \exp\left(10 q_t^j / \tau\right)}$$

where the parameter τ in $\{0.02, 0.25, 0.50, 0.75, 1\}$ directly relates to the intensity of the exploration motive. For $\tau = 0.02$, the organization picks with very high probability the alternative with the highest current propensity; this is an exploitative action. As τ increases, the choice probability shifts towards other alternatives and exploratory actions become more likely.

We argue that the evolution of propensities in (1) is not plausible for dynamic environments, because it is implicitly based on a cumulative accrual of evidence. When $\eta > 0$ and a shock displaces alternative i, the past outcomes for i become uninformative about the new value of p_t^i. However, Rule (1) keeps cumulating such stale evidence when computing the propensity for i. Moreover, since the weight for a new piece of evidence decreases as $1/(n_t^i + 1)$, the marginal impact of more recent information is decreasing; that is, the cumulative effect of past history tends to overwhelm fresh evidence. For instance, suppose that alternative i has had a long history of successes; if a negative shock makes p_i drop, the firm would take in a substantial streak of failures before its propensity q_t^i is brought back in line with the new value of p_i.

This bias may be partially corrected by a higher τ, because increasing exploration speeds up the alignment process between the propensity vector Q_t and the actual probabilities in P_t. However, this is inefficient because it takes ever longer streaks of experiments to overturn the cumulated past evidence. One of our goals is to demonstrate the advantages for an organization to shed away stale evidence in a turbulent environment.

Formally, the root of the problem in PL's setup is that the *marginal impact* of the last observation in Eq. (1) declines as $1/(n_t^i + 1)$. Among many different ways to correct this problem, an optimal choice should depend on η. However, the exact value of this parameter is unlikely to be known to the organization. Therefore, we opt for a simple rule that is robust to such lack of quantitative information about η. Its robustness comes from a built-in mechanism that modulates the intensity with which past evidence is shed away as a function of the degree η of turbulence in the environment.

We advocate two modifications to PL's learning model. Both refresh evidence endogenously. The first one deals with the possibility that the current choice may have been made unfavorable by a negative shock. When alternative i is chosen and $n_t^i \geq \bar{n}$, we split its past history into two segments of equal length: the first and the second half. (When n_t^i is odd, we include the median event in both histories.) We aim to drop from consideration the initial segment when a shock might have occurred and past evidence turned stale. To do so, we compute the average performances $\bar{R}_i^1$ and $\bar{R}_i^2$ over the first and the second segment, respectively. Then, with probability equal to $|\bar{R}_i^1 - \bar{R}_i^2|/2$, a *refresh* takes place: we delete the initial segment and recompute q_t^i accordingly. Since we only act when $n_t^i \geq \bar{n}$, the length of the past history after a deletion never goes below $\bar{n}/2$. For $\bar{n} \uparrow \infty$, we recover the model in PL. For demonstration purposes, in this paper we set $\bar{n} = 30$.

The second modification recognizes that alternatives that have not been tried in a long time may have been reset by a shock. In particular, whenever a refresh takes place, we reset the propensity for each alternative that has never been explored since the previous refresh to $1/2$.

In short, the first modification reduces the risk of staying with an alternative that has turned into a "false positive"; the second recovers forgone alternatives that might have changed into "false negatives". We refer to the model dealing only with false positives as M1, and to the full model as M2. We were surprised to discover how much M2 improves over M1 in a dynamic environment. Each of the values reported below is an average based on 5,000 simulations with different seeds.

3 The Stationary Environment

Our benchmark is the stationary environment, when $\eta = 0$. PL consider four indicators. *Performance* in PL is the cumulated value of rewards; for ease of comparison, we report the average performance $(\sum_{\tau=1}^{t} R_{\tau}^{*})/t$ per period, where R_t^* is the reward associated with the choice made at period t. *Knowledge* embodies the ability of Q_t to track P_t and is measured by $1 - \sum_i (p_t^i - q_t^i)^2$. The *Opinion* indicator $\sum_i (q_t^i - \bar{q}_t)^2$ is the sample variance of propensities; the higher it is, the more diverse the propensities and therefore the probabilities of choosing each alternative. Finally, the *Exploration* indicator computes the probability that the choice at time t is different from the choice at time $t - 1$.

PL report the values of these four indicators at $t = 500$. As it turns out, this horizon is too short to take into account the onset of the steady state and thus PL's analysis is limited to the short run. (They do not mention a rationale for this choice.) We replicate their short-run analysis at $t = 500$ and extend it to the long-run at $t = 5,000$. The short- and long-run values for PL are shown on the left-hand side of Table 1, respectively on the first and second line of each box. With a few exceptions (notably, when $\tau = 0.02$), differences in values between short- and long-run hover around 10 %. The working paper [4] provides a visual representation of the data, that we omit for brevity.

The left-hand side of Table 1 confirms and extends the short-run results in PL's Sect. 3.1. Exploratory behavior is increasing in τ, and the optimal level of search intensity τ is around 0.5. Except for $\tau = 0.02$, the long-run performance is about 10 % higher than PL's short-run estimate: since the search intensity never abates, this increase is not due to "cashing in" from reducing the searching efforts but instead stems from the long-run stationarity.

Knowledge and Opinion are similarly higher, as an immediate consequence of the larger cumulated number of experiences. The increase in Exploration is due to a little known property: in the short run, the softmax algorithm tends to ignore an alternative that has failed on the first few attempts, regardless of its actual probability of success. Any of such false negatives contributes towards making the algorithm focus on very few alternatives in its early stages. However, given enough time, the algorithm eventually returns to such alternatives and, if it finds them valuable, puts them back in the explorable basket. To gauge the extent of this effect, Table 2

Table 1 Performance, knowledge, opinions, and choices in the stationary environment

		PL	M1	M2
	τ	0.02 0.25 0.50 0.75 1.0	0.02 0.25 0.50 0.75 1.0	0.02 0.25 0.50 0.75 1.0
Performance	$t = 500$	0.48 0.52 0.56 0.54 0.50	0.53 0.55 0.56 0.53 0.50	0.54 0.55 0.55 0.51 0.47
	$t = 5{,}000$	0.49 0.58 0.61 0.59 0.55	0.59 0.61 0.61 0.59 0.54	0.61 0.61 0.60 0.55 0.50
Knowledge	$t = 500$	0.55 0.56 0.59 0.65 0.72	0.54 0.56 0.59 0.65 0.71	0.59 0.61 0.64 0.70 0.76
	$t = 5{,}000$	0.56 0.57 0.64 0.77 0.89	0.52 0.54 0.61 0.73 0.83	0.61 0.63 0.66 0.73 0.79
Opinion	$t = 500$	0.16 0.21 0.32 0.44 0.50	0.21 0.23 0.31 0.40 0.46	0.11 0.12 0.16 0.24 0.31
	$t = 5{,}000$	0.17 0.23 0.42 0.53 0.55	0.23 0.27 0.37 0.47 0.50	0.11 0.12 0.16 0.23 0.29
Exploration	$t = 500$	0.00 0.02 0.18 0.39 0.54	0.00 0.02 0.14 0.33 0.48	0.01 0.05 0.17 0.38 0.54
	$t = 5{,}000$	0.00 0.04 0.26 0.46 0.60	0.00 0.01 0.13 0.33 0.51	0.02 0.04 0.15 0.37 0.54

Table 2 Percentage of (almost) unexplored alternatives

	PL	M1	M2
τ	0.02 0.25 0.50 0.75 1.0	0.02 0.25 0.50 0.75 1.0	0.02 0.25 0.50 0.75 1.0
$t = 500$	0.89 0.86 0.79 0.69 0.60	0.83 0.80 0.74 0.65 0.57	0.81 0.78 0.71 0.62 0.51
$t = 5{,}000$	0.89 0.83 0.67 0.48 0.27	0.75 0.72 0.59 0.41 0.23	0.63 0.54 0.31 0.05 0.00

provides estimates for the percentage of alternatives that are explored less than $\bar{n}/2 = 15$ times in the whole period.

The rest of Table 1 provides data for our models M1 and M2, where old evidence may be discarded. One would expect PL to perform better in a stationary environment, because P_t is constant over time and thus evidence never gets stale. However, by forgetting stale evidence, both M1 and M2 refresh propensities and have an endogenous bias towards more search. Such bias overcomes the "false negatives" trap of the softmax algorithm and makes their performance competitive with (and often marginally better than) PL. In particular, both M1 and M2 achieve their superior performance with a lower level for the Exploration indicator: compared to PL, they are less likely to switch the current choice.

Instead of PL's five-points grid, we computed the optimal search intensity over a finer 100-points grid and found the following optimal values: $\tau = 0.56$ (0.48) for PL when $t = 500$ ($t = 5{,}000$, respectively); $\tau = 0.45$ (0.36) for M1; and $\tau = 0.40$ (0.24) for M2. The sharp reduction in the optimal search intensity from PL to M1 to M2 stems from their search bias. Within each model, the optimal τ decreases when going from the short- to the long-run because steady state learning is more effective.

We summarise our comparative evaluation of the three learning rules. The search intensity τ is not easy to tune in practice, but our models are more robust: they deliver a tighter range across different values of τ. This comes with less switches in choice and tighter opinions (for the same level of τ), and an overall comparable performance. Thus, although the three learning models are roughly comparable in a static environment, ours are more robust.

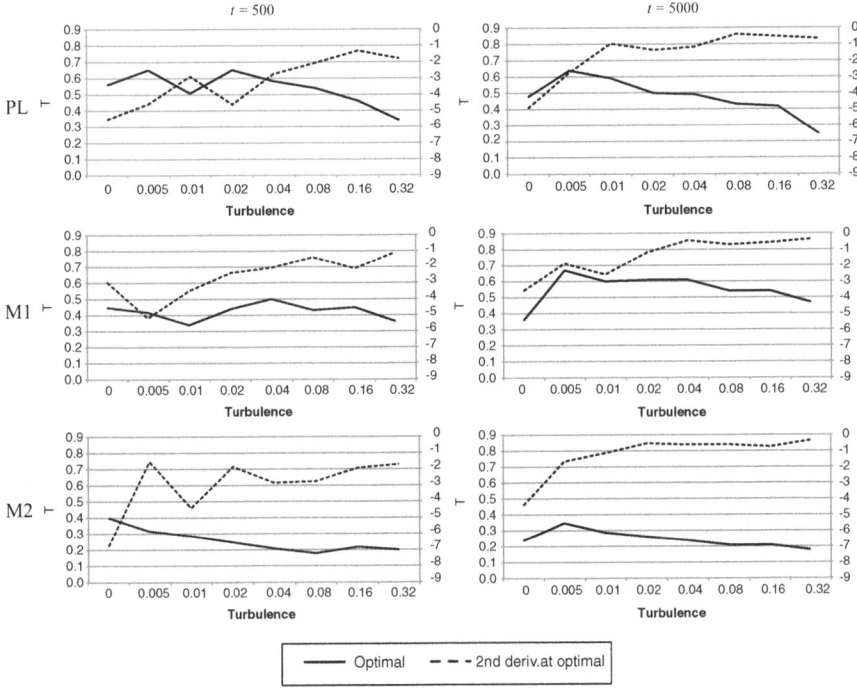

Fig. 1 Optimal exploration strategy across turbulence levels

4 The Dynamic Environment

In a dynamic environment, turbulence is represented by the probability $\eta > 0$ that in each period a shock resets the actual probabilities in P_t. Following PL, we consider $\eta = 0.005 \times 2^k$ for $k = 0, 1, \ldots, 6$. The main result in PL is that the optimal level of search intensity has an inverse U-shaped form that is right skewed. We found that this statement must be qualified as follows.

PL derive the curve by "fitting a third order polynomial to the results" (p. 593), but no details are provided and the available points are just five. Therefore, we opted for a brute force approach and did and extensive search over $[0.02, 2.00]$ using a grid with mesh 0.01. Figure 1 illustrates the results, reporting data for $t = 500$ and $t = 5,000$ on the left- and right-hand side, respectively.

Let us begin with the long run ($t = 5,000$), as represented on the right-hand side of Fig. 1. For $\eta > 0$, the optimal search intensity is actually decreasing in the turbulence level. The inverse U-shaped form is a visual artefact created by the inclusion of the first datapoint ($\eta = 0$) corresponding to zero turbulence. In a dynamic environment, an organisation with a sufficiently long horizon has an optimal search intensity that is decreasing in the level of turbulence.

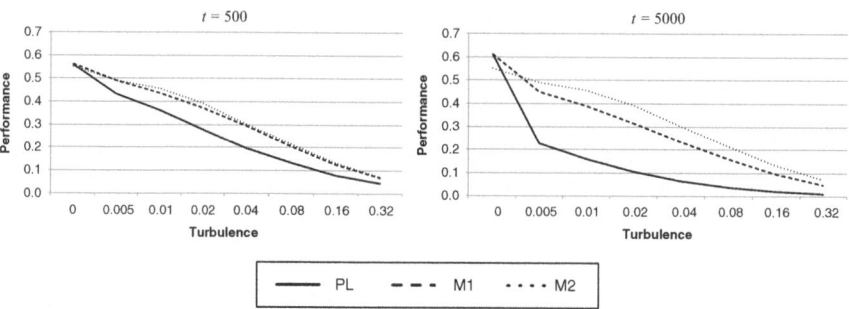

Fig. 2 Performance across turbulence levels ($\tau = 0.5$)

Consider now the short run ($t = 500$). We find a dip at $\tau = 0.01$ for PL, but this might be due to noise. On the other hand, the optimal search intensity for M1 stays pretty flat, and for M2 is decreasing overall. We argue below that M2 is superior to PL; thus, for an organisation with an appropriate learning model and a short run horizon, the optimal search intensity is *decreasing* in the level of turbulence. We conclude that, under an appropriate model specification, turbulence has a systematic negative effect on the optimal search intensity: the inverse U-shaped form claimed by PL is not an accurate depiction of this result.

PL discuss how the value of the second derivative of the performance at the optimal τ can be used as a proxy for the intensity of the tradeoff between exploration and exploitation. While generally negative, the closer to zero, the flatter the curve $f(\tau)$; and hence the less important pinning down the right τ is. Lack of details in PL prevented us from replicating their work, so we decided to compute our approximation to the second derivative in two steps. First, for each point τ on our grid, we computed the second-order central difference $D(\tau) = [f(\tau + 0.01) - 2f(\tau) + f(\tau - 0.01)]/h^2$. Second, we performed a simple smoothing by replacing $D(\tau)$ with the weighted mean

$$\overline{D}(\tau) = \frac{D(\tau - 0.02) + 2D(\tau - 0.01) + 4D(\tau) + 2D(\tau + 0.01) + D(\tau + 0.02)}{10}$$

The graph for the (approximated) second derivative is superimposed as a dashed line on the panels in Fig. 1. After cautioning the reader not to put much weight on the first datapoint ($\eta = 0$), we find that in most cases the second derivative is increasing in the turbulence level, confirming PL's claim that pinpointing the optimal τ matters less to performance when turbulence is higher.

Coming to performance, we were puzzled by the contrast between PL's extensive discussion of it for the stationary environment ($\eta = 0$) and the complete lack of data for $\eta > 0$. A primary element in evaluating the plausibility of the learning rule under turbulence should be its performance. Figure 2 provide a visual representation of the data for $\tau = 0.5$. (The working paper provides tables with the numerical values for this figure as well as for the following ones.) Here, as in PL, we leave the search intensity τ constant. Alternatively, one might consider the optimal performance using the best search intensity for each η. We report the outcome of this exercise in the working paper: we found qualitatively similar results that are even more favourable to the claim we advance below. Hence, fixing $\tau = 0.5$ avoids biasing the graphs against PL.

Except when $\eta = 0$, the performance for M1 and M2 is consistently and significantly better than for PL over both horizons. In the long run, the degradation in performance for PL is much stronger and, if one ignores the data point for $\eta = 0$, fairly disastrous: PL scores about 20 % when turbulence is minimal ($\eta = 0.005$) and drops to virtually 0 %—equivalent to random choice—under intense turbulence ($\eta = 0.32$). It is hard to claim that PL's rule captures effective learning in a turbulent environment.

To the contrary, both of our models deal with intense turbulence reasonably well. The decline in performance when η increases is not as abrupt as PL and, even under intense turbulence, they manage to rake up a performance that is small but significantly higher than the 0 % associated with random choice. Moreover, by explicitly dealing with the foregone alternatives that shocks might have turned into false negatives, M2 performs significantly better than M1 in the long run. Therefore, when a shock is deemed to have occurred, one should not only drop evidence about the (potentially) false positive as in M1, but also about the (potentially) false negatives as in M2. This is worth pointing out because many studies about the representativeness heuristic suggest that people are less prone to review evidence about false negatives than about false positives. The main conclusion is that shedding stale evidence makes the search process in a dynamic environment perform better as well as exhibit resilience to turbulence.

PL convincingly argue that turbulence erodes performance by two effects: it alters the future value of existing knowledge and reduces the payoff from efforts to generate new knowledge. To disentangle these two effects, they use a differences-in-differences analysis assuming a search intensity $\tau = 0.5$. (See PL for details.) Their approach separately estimates the accretion of new knowledge and the erosion of existing knowledge for different levels of turbulence. These two effects jointly determine the net change in knowledge. We replicated their short-run analysis ($t = 500$), and extended it to the long-run ($t = 5,000$) using propensities at $t = 4,000$ and $t = 5,000$. As before, the choice $\tau = 0.5$ fits PL better than our models; but, again, we redrew the graphs using the optimal value of τ for each turbulence level, and found no qualitative differences. Using PL's setting for ease of comparability, the results are shown in Fig. 3.

We found again that the details in some of PL's statements need amendments. Looking at the short-run, all models exhibit the same behaviour; namely, both accretion and erosion have an inverse U-shaped form and the net effect on knowledge is overall positive across all levels of turbulence. The size of the two effects, however, is quite different: in PL none of the two effects brings about a change greater than 8 % in absolute, while in M1 and M2 this can go as high as 14 %. The vertical dilation in the graphs as we move downwards from PL to M2

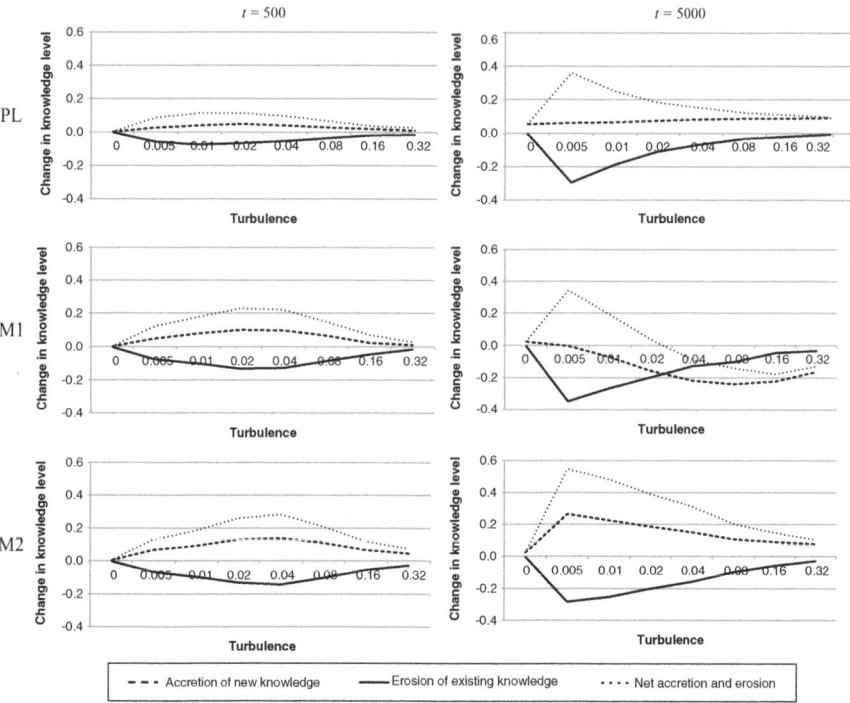

Fig. 3 Knowledge accretion and erosion across turbulence levels ($\tau = 0.5$)

on either side of Fig. 3 is apparent. Shedding evidence magnifies both the positive accretion effect and the negative erosion effect.

Over the long run, the two effects change shape for all models, after discarding the insignificant datapoint at $\eta = 0$. Conforming to intuition, one would expect knowledge accretion and knowledge erosion to be respectively decreasing and increasing in turbulence. This occurs only for M2, while PL and M1 match the pattern for knowledge accretion only partially. PL exhibits knowledge accretion that is increasing in turbulence. M1's knowledge accretion is decreasing over most of the range, but eventually starts climbing up generating a U-shape. Given that M2 is superior in what regards both performance and the net effect on knowledge, it is reassuring to see that the pattern of its knowledge accretion effect matches intuition.

Our last batch of work replicates and extends PL's Fig. 6 reporting the accuracy of knowledge, the strength of opinions, and the probability of exploration at $\tau = 0.5$ in Fig. 4. Over the short run, the three models exhibits the same qualitative shapes for the three indicators and these are consistent with intuition. With respect to turbulence levels, knowledge is decreasing, strength of opinions is decreasing, and probability of switching choice is increasing.

Moving to the long run reveals a few hidden patterns. First, the knowledge indicator goes almost flat for PL, suggesting that the knowledge generated within

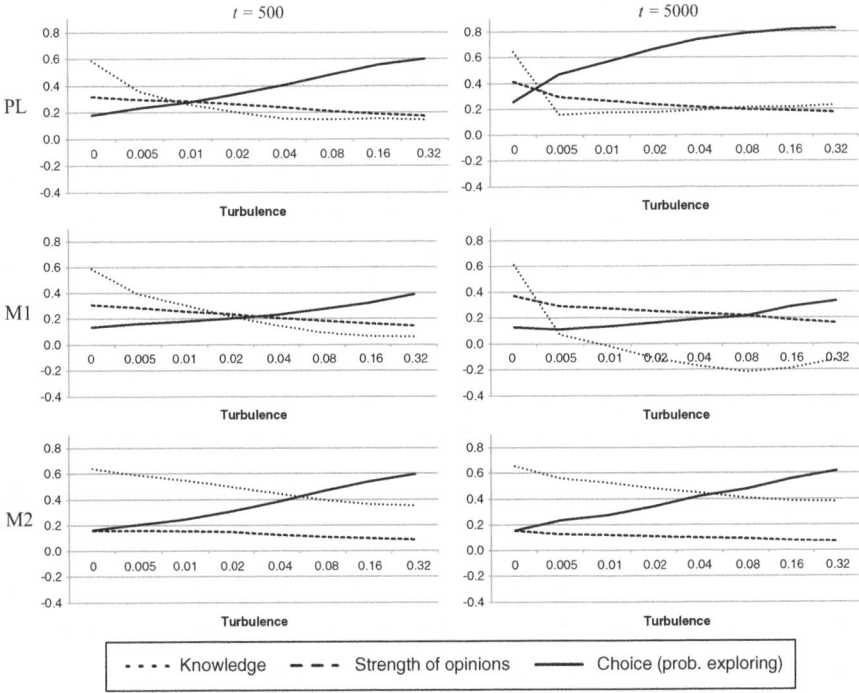

Fig. 4 Knowledge, opinions, and choices across turbulence levels ($\tau = 0.5$)

this model in the long run is unaffected by the level of turbulence. (Differently put, once we enter the steady state, the level of turbulence has a negligible effect on knowledge.) With little variation in opinions, PL ends up with very similar propensities across all alternatives and, accordingly, the probability of switching becomes much bigger: in practice, PL ends up being close to (randomly) wandering across alternatives. M1 generates even less knowledge in the long run, but its strength of opinions is bigger: in other words, propensities are more polarised (which helps focusing choice and reduces the probability of switching) but on the wrong alternatives (which adversely affects knowledge).

Finally, M2 is very effective in the long run: its knowledge indicator is small and decreasing with respect to turbulence, because in a dynamic environment it is ineffective to strive for high levels of knowledge. Keeping knowledge small ("pack light") allows opinions to change swiftly and track shocks accurately; hence, their strength resists homogenisation and stays around 0.4 even when turbulence is intense. Finally, the probability of switching choice increases less than PL and more than M1: in other words, the action bias of M2 is intermediate. This is necessary to balance two opposing effects: the risk of wandering choices (as in PL) against the possibility that exploration cannot keep with the flow of incoming shocks. Notably enough, M2 achieves this balance endogenously: our models have not been calibrated for maximum performance.

5 Conclusions

We revisit a recent study by Posen and Levinthal [6] about learning under turbulence. We claim that their analysis overlooks the long run and posits a learning model that puts too much weight on stale evidence. This leads us to suggest two learning models that incorporate an endogenous mechanism to spot and shed away obsolete evidence. M1 deals only with the possibility that some shock may have made the current choice a false positive, while M2 adds a concern for foregone alternatives that may have become false negatives. PL is nested into M1, and M1 is nested into M2.

The comparative analysis shows that M2 offers a significantly superior performance, making PL an implausible candidate for an effective learning model. Even under intense turbulence, its ability to "pack light evidence" makes it properly responsive to shocks, and allows it to deliver a performance that is both robust and resilient. We believe that clarifying the importance of giving up on obsolete evidence is the major contribution of this paper.

Finally, we carry out a comparative analysis for several claims in Posen and Levinthal [6], both over the short and run long run and across the three models. While their main insights survive, we find and point out which qualifications are needed for their validity. In particular, some of the (somewhat unintuitive) non-monotone relationships they discover using PL in the short run disappear when the analysis is carried out in the long run using M2.

References

1. Berry D, Fristedt B (1985) Bandit problems. Chapman and Hall, London
2. Cesa-Bianchi N, Lugosi G (2006) Prediction, learning, and games. Cambridge University Press, New York
3. Duffy J (2006) Agent-based models and human subject experiments. In: Tesfatsion L, Judd KL (eds) Handbook of computational economics, vol 2. North-Holland, Amsterdam/New York, pp 949–1011
4. LiCalzi M, Marchiori D (2013) Pack light on the move: exploitation and exploration in a dynamic environment. Working Paper 4/2013, Department of Management, Università Ca' Foscari Venezia,
5. March JG (1991) Exploration and exploitation in organizational learning. Organ Sci 1:71–87
6. Posen HE, Levinthal DA (2012) Chasing a moving target: exploitation and exploration in dynamic environments. Manag Sci 58:587–601
7. Sutton RS, Barto AG (1998) Reinforcement learning: an introduction. The MIT University Press, Cambridge

Part VI
Networks

An Agent-Based Model of Access Uptake on a High-Speed Broadband Platform

Fernando Beltrán and Farhaan Mirza

Abstract We model the access uptake on a newly built high-speed fibre-to-the-home (FTTH) broadband network using a computational Agent Based Model (ABM). Two cases illustrate the model analysed in this paper: the Ultra-Fast Broadband (UFB) Network in New Zealand (NZ) and the National Broadband Network (NBN) in Australia. Common learnings of both projects are used in our model to describe and analyse the uptake of fibre connections to households and businesses. By design network operation is decoupled from service provision and the platform is open-access, meaning any provider can operate end-user services. In our model a high-speed broadband network is regarded as a two-sided platform that accommodates both end-users and service providers, creating the conditions for the two sides to exploit mutual network effects. Results show that the greater the number of users (end-users or providers) on one side, the more the number of users (provider or end-users) on the opposite side grows. Providing free connections and raising consumer awareness is a means for driving consumer uptake. Scenario based analysis allows us to investigate the magnitude of network effects' on the fibre connection uptake.

1 Introduction

Recent national broadband initiatives have led to the construction of country-wide, fibre-based broadband networks. Countries such as Australia, NZ and Singapore have adopted a high-speed, FTTH network model where public funds are invested (with or without participation of private partners) and an open access operation

F. Beltrán (✉) · F. Mirza
University of Auckland, 12 Grafton Rd, Auckland, New Zealand
e-mail: f.beltran@auckland.ac.nz; farhaan@mirzabros.com

is adopted. This paper models a high-speed, open access broadband network as a two-sided platform which comprises of two markets: an *access market* and a *content market*. The scope of the *access market* involves consumers achieving fibre connectivity from their network provider known as Local Fibre Company (LFC). The content market involves consumers subscribing to a Retail Service Provider (RSP) for actual retail services and products such as Internet access, voice service, and video (broadcast TV or on-demand). In this paper we focus on the access market and exclude issues of content market because the platform is still under construction. Focusing on consumer uptake helps understand salient aspects of FTTH growth when both, attractiveness of the platform to end-users and incentives for RSP participation are considered through the network effect approach. New high-speed broadband network build-up presents us with a timely unique opportunity to analyse the issues and drivers for uptake and growth as the markets take shape and evolve. Our approach builds upon [1] where both access market and content market are modelled as a developing two-sided market platform. We use an Agent-based Model (ABM) to simulate a range of scenarios that illustrate how the broadband uptake rate is affected by varying factors.

The rest of the paper develops as follows. In Sect. 2 we present the market structure of the FTTH platform and relate it with the theory of two-sided platforms, followed by highlighting key features of agent-based modelling. In Sect. 3 we present the access market model and provide details on configuration of simulation scenarios. We describe how the agents will interact in the model including consumer decision making process. Section 4 presents the key findings followed by conclusions in Sect. 5.

2 A High-Speed Broadband Market as a Two-Sided Platform

The new high-speed broadband structure will see a vertical separation of ownership and operation with LFCs being owners and RSPs being providers of services. LFCs will own and operate the lower layers of the network which will be geographically exclusive providing them a monopoly on the franchised area. This structure (Fig. 1) has introduced a vertical separation (lower layers structurally separated from upper layers), giving birth to an open-access platform. The main task of each LFC is to install fibre connections to their consumers in their regions and maintain fibre lines. Higher benefit will be achieved if LFCs manage to quickly implement the network, as it would satisfy the prerequisite for RSPs to start retail services. The retail services can only be sold by the RSPs to consumers. RSPs will purchase layer 2 services at regulated wholesale prices from the LFCs. The RSPs are expected to develop business models that include high-speed broadband, phone line, IPTV and video services over the platform.

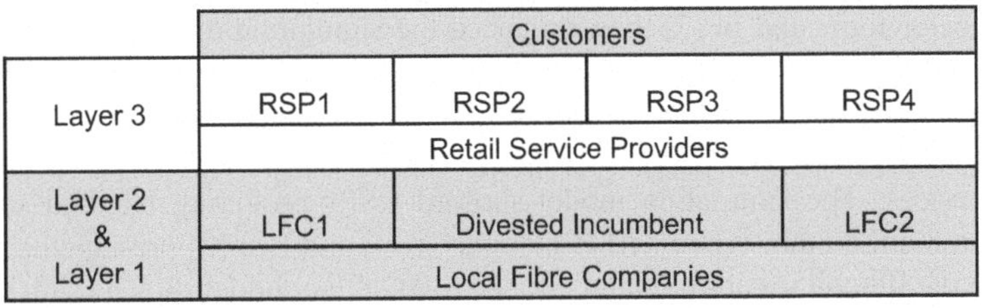

Layer 3	Customers			
	RSP1	RSP2	RSP3	RSP4
	Retail Service Providers			
Layer 2 & Layer 1	LFC1	Divested Incumbent		LFC2
	Local Fibre Companies			

Fig. 1 High-speed broadband market structure

2.1 The Broadband Platform as a Two-Sided Market

Founding work on the economics of two-sided platforms is found in Rochet and Tirole [2, 3] who introduced the term two-sided market. In a traditional value chain diagram, value moves from left to right: to the left of the company is cost; to the right is revenue whereas in two-sided networks, cost and revenue are both to the left and the right, because the platform has a distinct group of users on each side [4].

An open-access, high-speed broadband network represents a scenario of a two-sided platform [1]. A network provider implements a fibre-based network which can be regarded as a platform. On one hand the platform sells wholesale services to RSPs, whereas on the other it offers fibre connections to consumers. It is common in two-sided markets for one of the user groups to be subsidised. For example the Yellow Pages is usually free of cost to consumers, but advertisers pay to get a featured advertisement. Usually job seekers access job portals for free and employers need to pay to advertise. If these platforms reversed their approach, their network probably would not exist. In the case of FTTH network consumer access is being partially subsidised to encourage consumer participation.

2.2 Agent-Based Approach

The high-speed broadband market comprises multiple stakeholder groups, each with their own self-serving objectives. The stakeholder groups of LFCs, RSPs and consumers can be represented as heterogeneous agents in an ABM, each behaving according to their own preferences. Our ABM demonstrates how some market activities can be generated by the endogenously evolving interactions among such boundedly-rational stakeholders over time [5].

The use of ABM is not peculiar to the analysis of telecommunication markets. It is an effective methodology to tackle dynamics of telecommunication markets that involve: changing technologies, products, and regulatory reforms. Among the work conducted on telecommunication markets using ABM following is a sample of relevant references.

Beltrán and Roggendorf [6] used ABM to create an auction-based pricing scheme to facilitate network resource distribution negotiations for the analysis of bidding behaviours and in [7] they enhanced the simulation model by introducing richer strategies. Baryshnikov, Borger, Lee, and Saleh [8] created consumer and service provider agents assigned with utility based preference scores. The model included two types of RSP agents, RSPs providing bundled services vs. undiversified services. The simulation model showed RSPs providing bundled services outperformed the other type of RSP. Douglas, Lee, and Lee [9] presented a model in which the iPhone was introduced into the market, the model showed satisfactory reproduction of historical data but failed to predict exact market share in the future. Zheng, Jin, and Zhang [10] explored the effects of regulation with a duopoly mobile market and found that the duopoly market operated more efficiently with regulatory interventions. Diedrich and Beltrán [11] leveraged ABM to compare traffic discrimination policies. The model varied policy and competition scenarios. The results found that the content providers performed best when network neutrality is imposed; while network providers and consumers may benefit from traffic discrimination under certain circumstances. In the following text we present the ABM for the high-speed broadband platform access market.

3 Participation of RSPs and Consumers in the Access Market

Beltrán [1] used two-sided market theory to explain the presence of cross-network effects on an open access broadband platform. Cross-network effects embody the interdependencies between the two sides. One side, the RSP side, purchases wholesale services from the platform operator. The other side, the end-user side, is split in two groups: residential (R) customers and business (B) customers.

3.1 Access Market Model

C_{RSP} represents the wholesale cost to a RSP to get services from the platform. The number of residential and business consumers with active subscriptions from the RSP are given by: n_{AR} and n_{AB}. P_R and P_B are the wholesale rates payable to the platform per individual residential or business consumer Therefore we can define the C_{RSP} as below:

$$C_{RSP} = n_{AR}P_R + n_{BR}P_B \qquad (1)$$

Regulated wholesale prices are charged based on individual connections (each of n_{AR} and n_{AB}). This means the LFC owned platform obtains revenue C_{RSP} from the RSP side based on the number of consumers it will manage to subscribe.

An RSP's utility function would consider the positive effect of the presence of residential (n_{FR}) and business consumers (n_{FB}) with potential fibre connectivity (passing fibre line on the street) less the cost of purchasing wholesale services from the platform. In other words, the market becomes increasingly attractive to the RSP as *passing fibre* becomes available to homes and businesses. On a first approach business fibre consumers (n_{FB}) are excluded mainly because their associated connection costs are not clearly defined at this early stage of implementation (at least in the international cases inspiring the present model). The expenses that need to be paid to the platform by a RSP are represented by P_{RSP}, which is equal to C_{RSP}. Thus, if n_{FR} is the number of residential passing fibre consumers – who may potentially activate a fibre connection with a RSP, the utility of a RSP, U_{RSP} can be written as, Eq. (2). U_{RSP} increases as n_{FR} increases over time (t) and α_{RSP} measures the effect of each consumer's platform presence perceived by the RSP.

$$U_{RSP}(t) = n_{FR}(t) - \alpha_{RSP} n_{FR} \qquad (2)$$

The other side of the access market involves consumers. A residential consumer's utility function is expressed as:

$$U_R(t) = n_{RSP}(t)\alpha_R \qquad (3)$$

where α_R measures the effect of each of the n_{RSP} RSPs presence connected to the platform, which is perceived by a representative consumer. An additional assumption is that residential users may or may not pay for their fibre connection to home; this will depend on the subsidy terms defined by LFCs and the government.

The cross network issues described above were setup as an ABM [12] for the NZ case in order to highlight the impact of these cross network effects between RSPs and consumer groups. In this article we introduce further improvements to the ABM of [12].

We conducted an empirical study in NZ by engaging with leading platform operator (Chorus), RSP (Snap Internet) and consumers using a dominant imperialist multi-methodological design – one method or methodology as the main approach with contribution(s) from other(s) [13]. In our case the dominant method is qualitative data from 15 consumer interviews. The contributions include: firstly research study data from Chorus that included a large sample of 132 qualitative interviews followed by an online quantitative survey that totalled 1,009 respondents. Secondly broadband customers sample data from Snap's Customer relationship management (CRM) system – which included information including the type of broadband products consumers are using, costs, location, and why the consumer decided to use Snap. The aim of the interviews was to acquire detailed information about the consumer perceptions of participating in the access and content market of a high-speed FTTH platform. The results would complement the secondary data from Chorus and Snap.

In this article we are leveraging the results obtained from this empirical study, therefore only describing a brief summary of relevant details for the ABM. We found the main driver for consumers to participate in the FTTH access market was perceived platform's reliability and consistently fast connection. The barrier for participation was consumer contentment with the inferior ADSL or alike alternatives, however in the Australian NBN initiative the plan is to eventually make the old technologies obsolete. The deciding factors which promotes or withdraws consumer participation in the access market were firstly, *start up costs* which include cost of connection. Secondly, *awareness* regarding the benefits of fibre technology and its products. We incorporate these deciding factors in our ABM.

The price of the connection to the street is often met by the government or the platform operators. However the connection into the house may or may not be subsidised. In NZ the connection from curb to home is free until 2015, however not a very small number of potential consumers have taken benefit of such an offer. Chorus revealed only a 1.7 % uptake of 80,299 end users able to connect [14]. This report echoes our findings related to awareness being an important driver for consumer participation.

3.2 Setting Up Simulation Scenarios

The model simulates consumer uptake of a newly developed open-access, high speed broadband platform. End-users find it attractive to connect to the network when they find that a large number of RSPs operate on the platform, a fact that enhances consumers' expectations of service.

RSP and consumer agents will be assigned a unique utility score by the simulation model and this will influence how they participate in this two-sided platform, based on Eqs. (2) and (3). The values for the α_{RSP} and α_R are randomly generated from probability distributions. RSPs will slowly appear into the market and will eventually reach saturation. New consumers, on the other-hand, keep subscribing to platform services as the connection becomes available. For the RSPs the larger number of consumers adopting fibre, the more encouraged they will be to enter the market. RSP agents maintain a *when-to-market* attribute, this specifies when a RSP should enter the market. This is when the U_{RSP} is positive. The consumer on the other hand maintains a *when-to-subscribe* attribute. These attributes vary based on the scenario settings and the probability distribution of the network effect parameter. We set up the following scenario groups:

1. Strength of cross network effects are *high* for consumers and RSPs.
2. Strength of cross network effects are *high* for consumers and *low* for RSPs.
3. Strength of cross network effects are *low* for consumers and *high* for RSPs.
4. Strength of cross network effects are *low* for consumers and RSPs.

3.3 Consumer Awareness

The consumer can become aware by multiple means. For this model we setup the following three drivers. Firstly, the platform operating LFC may run marketing campaigns causing an increase in awareness in a given percentage subset of the population. For this model we configured these awareness campaigns to occur every 6 months reaching 5 % of consumers. Secondly the consumer becomes more aware when the number of RSPs in the market increases beyond the consumer's own perceived utility. Lastly we found from the interviews and industry data that family, friends, and word of mouth advertising is a trusted way for consumers to become aware. Therefore we integrated a friend circle creator model [15] into this ABM. This allows the consumer agents to become informed via their friend circle; in the case of this ABM the awareness goes up for the consumer when a majority of his friends subscribe to RSPs. Awareness score is maintained for each consumer and a score of 5 is considered to be high awareness, and anything lower is low awareness.

The scenario groups above are further explored by varying connection subsidisation within the market and consumer awareness settings. As a result each of the scenario groups 1–4 specified are further explored by the scenarios A–D below, producing a total of 16 scenarios.

A. Partial subsidy and low awareness
B. Partial subsidy and high awareness
C. Full subsidy and low awareness
D. Full subsidy and high awareness

The agents are: a single broadband platform operator, a number of RSPs and a large number of residential consumers. The following text will describe each agents behaviour in the model.

The platform operator is tasked to implement fibre, which involves adjusting the consumer's status from *no fibre* to *passing fibre*. The speed of implementation is scaled relatively to the actual implementation – which is around 5–7 years in the international FTTH cases. In the reported results (Fig. 4) we configured the simulation for 250 ticks for the platform operator achieving 2,500 homes with *passing fibre* status. In this term all of the consumers will have fibre installed at their premises (i.e. $n_{FR} = 2,500$), making the implementation speed approximately ten houses per day. The RSPs behaviour is simple – which is to become active in the market when U_{RSP} becomes positive, as explained above.

3.4 Consumer Agent Decision Making Process

Consumer agents maintain a connection status attribute. The connection status is either – *no fibre*, *passing fibre* or *subscribed to RSP*. The model creates all consumer agents with a *no fibre* status. The fibre eventually reaches consumer's curb based on

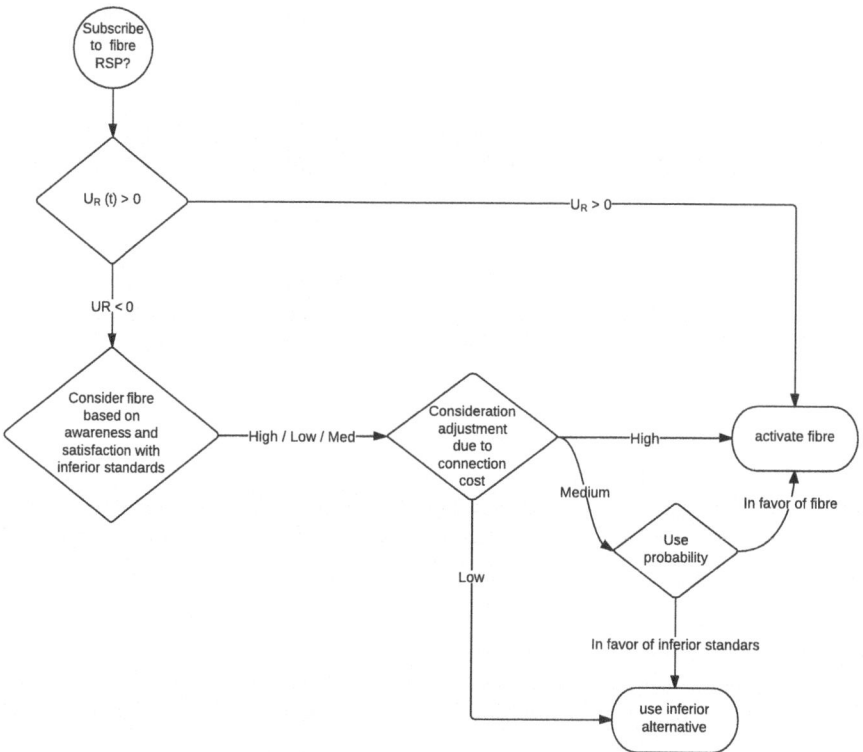

Fig. 2 Consumer decision making process for considering subscribing to a fibre retailer

the speed at which the platform agent is laying fibre, this changes the consumer status to *passing fibre*. This is when the consumer becomes eligible to subscribe to a RSP in order to benefit from the fibre based services. The decision making process for a consumer transitioning eventually to activating high-speed services is shown in Fig. 2.

When the consumer's street receives passing fibre, the LFC usually informs the consumer regarding their construction schedule and identify which RSPs are operating on their platform. This is when the process presented in Fig. 2 starts, whereby the consumer will check if $U_R(t)$ is positive. If $U_R(t)$ is positive, the consumer agent will activate their fibre connection with a RSP. Incases when the $U_R(t)$ isn't positive then the consumer agent defers its reconsideration till its awareness score increases over a configured threshold. This action is triggered endogenously at the time when each consumer's awareness becomes high, in this ABM its set to a score of 5. Consumers are also rationally bounded as they may or may not become aware based either via their friends or through LFC's marketing impact on the environment of the model.

While reconsidering, the consumer agent obtains a score from the matrix shown in Fig. 3, this value is based on a combination of present awareness and

	Dissatisfaction with existing connection	
	Low	High
Awareness — High	2	3
Awareness — Low	1	2

Fig. 3 Matrix for determining a score for consumer reconsidering fibre

dissatisfaction with existing connection. The consumer agent upon creation is profiled to have a certain type of satisfaction score with the alternative broadband technology to fibre. The consumer with high awareness of fibre services and with an unsatisfactory alternative connection, will be most likely to transition to fibre. This score is further scaled – either higher or lower depending on the present subsidy conditions in the market. As a result, the consumer either activates fibre or decides to remain on the inferior alternative, such as ADSL.

The access market model described above is a simplification of the many complex issues (political as well as economic) surrounding the build-up of a government-funded FTTH network. Fibre-access uptake in this kind of subsidised environment presents itself with issues not found in full private network expansion.

4 Simulation Results

Cumulative results are collected to appreciate how uptake rate is affected by the combination of cross-network effects, consumer awareness and connection pricing as shown in Fig. 4. Each plot displays the number of consumers connecting to the platform and subscribing to a RSP as a function of time. The inclining straight dashed line shows the number of households the fibre is passing. The solid black line with varying values shows the number of consumers who *subscribed to a RSP*. Underneath each plot the percentage of consumers that subscribed to a RSP is given along with number of active RSPs in the market at the end of each simulation run. The plots shown in Fig. 4 are averages of running the simulation a number of times.

The platform best outcome from the cross network effects perspective is found in scenario group 1 (high utilities) and the worst outcome in scenario group 4 (low utilities). Scenario 1D is the best platform outcome as expected because the consumer awareness is high, connection costs are subsidised, and the strength of cross network effects is high for RSPs and consumer sides of the platform. The opposite applies for scenario 4A. The conditions in scenario 1D manage to subscribe 46 % of consumers to a RSP and 11 RSPs become active in the market. The worst

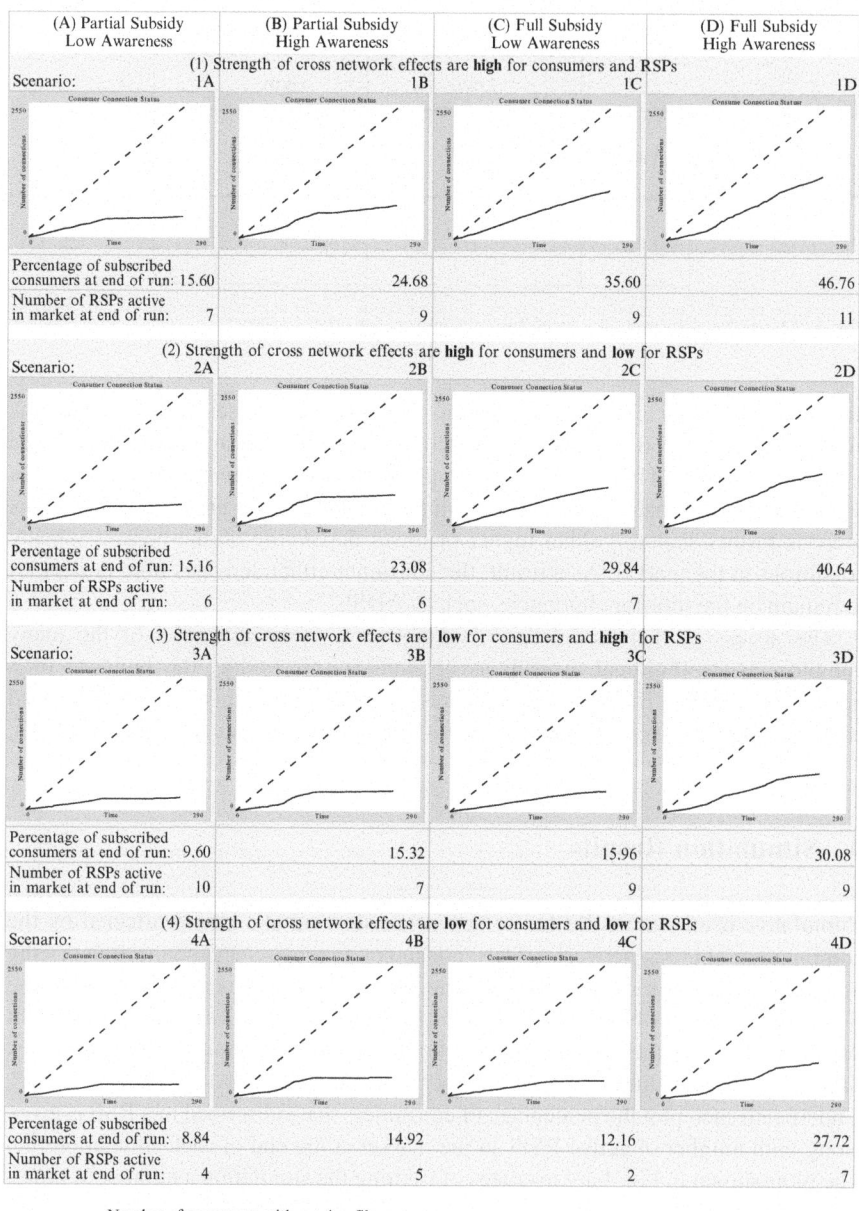

Fig. 4 Simulation results of varying scenarios

conditions of 4A could only subscribe 9 % of consumers to a RSP and 4 RSPs became active in the market. The results of the best and worse outcomes are as expected.

Scenario group 2 is closest to the best outcome scenario group 1 and scenario 3 is closer to the worst outcome scenario group 4, emphasising effects of high U_R having a greater effect on the platform than U_{RSP}. For example if we compare the percentages of consumers that subscribed to a RSP between scenario 2D (41 %) and scenario 3D (30 %), we can see that the subscribers 2D score significantly higher despite the active number of RSPs were four. This shows that the consumers took advantage of free connections because they were aware of the benefits of the high-speed broadband platform.

In scenario groups C and D – the connections were fully subsidised; whereas the scenario groups A and B provided partial subsidy, which expired in mid-implementation term. The curvature in the plots show how removing the subsidy limits the consumer uptake in the platform. It is common in two-sided markets for one of the user groups to be subsidised. The simulation shows the negative effects of not providing the subsidy.

If we compare the end of run values for percentage of subscribed consumers in scenario groups 3 and 4, we can see that there is a little difference. For example scenario 3B (15.32 %) versus scenario 4B (14.92 %). The advantage of scenario group 3 is such that the rate of consumer subscriptions were higher than scenario group 4 from the start, which is much more beneficial as the platform would be generating revenue from an early phase of implementation. This is because more RSPs became active in the early stages of implementation.

The friend recommendations helped uplifting the uptake percentages especially when cross network effects were low and subsidisation was partial. Friend recommendations increased the uptake ranging between 2–4 %.

5 Conclusion

The platform embodies an architecture that separates network services provision from end-user (retail) service provision. It operates under a set of rules which are in place by design (and regulatory intervention) whereby the LFC is prohibited from selling services directly to consumers. RSPs deal with the platform to acquire wholesale services which are used as inputs to their end-user services. We postulate the existence of cross network effects on the platform whereby end-users represented by consumer agents and RSPs represented by provider agents find themselves mutually attracted with an agent type's attraction level increasing, the larger the number of agents of the other agent type in the platform.

A high consumer utility assures the access market's performance, regardless of RSPs having a high or a low utility. Additionally the results show that having large number of first mover RSPs allows the platform to keep generating revenue, even though the consumer utility may remain low. The deciding factors for consumers in transitioning to the high-speed platform include awareness, subsidised connection to home, dissatisfaction with their existing broadband connection, and friends' influence. The simulation scenarios display a wide variation of these factors to demonstrate how they affect consumer uptake. Critical success factors for a successful FTTH market establishment includes evaluating subsidy considerations along with upliftment of consumer awareness. This will prevent delays in consumer uptake which will benefit the overall market by attracting revenue from early stages.

Our work contributes to the increasing literature on agent-based modelling of market performance whose main features include: modelling the FTTH access market structure, understanding the access market as a two-sided platform, and the computational ABM for a scenario-based analysis.

Acknowledgements The authors want to acknowledge the support that Centre of Digital Enterprise (CODE) at University of Auckland – Faculty of Business and Economics has provided to this work.

References

1. Beltran F (2012) Using the economics of platforms to understand the broadband-based market formation in the New Zealand Ultra-Fast Broadband Network. Telecommun Policy 36(9):724–735
2. Rochet JC, Tirole J (2002) Cooperation among competitors: some economics of payment card associations. Rand J econ 549–570
3. Rochet JC, Tirole J (2003) Platform competition in two-sided markets. J Eur Econ Assoc 1(4):990–1029
4. Eisenmann T, Parker G, Van Alstyne MW (2006) Strategies for two-sided markets. Harv Bus Rev 84(10):92
5. Windrum P, Fagiolo G, Moneta A (2007) Empirical validation of agent-based models: alternatives and prospects. J Artif Soc Soc Simul 10(2):8
6. Beltran F, Roggendorf M (2005) A simulation model for the dynamic allocation of network resources in a competitive wireless scenario. In: Mobility Aware Technologies and Applications. Springer, Berlin/Heidelberg, pp 54–64
7. Beltran F, Roggendorf M (2006) A simulation-based approach to bidding strategies for network resources in competitive wireless networks. In: Performability Has its Price. Springer, Berlin/Heidelberg, pp 37–48
8. Baryshnikov Y, et al (2008) Modeling market dynamics in competitive communication consumer markets. Bell Labs Tech J 13(2):193–208
9. Douglas CC, Lee H, Lee W (2011) A computational agent-based modeling approach for competitive wireless service market. In: 2011 international conference on information science and applications (ICISA), Jeju Island. IEEE
10. Zheng W, Jin X, Zhang B (2011) The simulation study based on multi-agents of mobile market. Adv Mater Res 143:433–438
11. Diedrich S, Beltrán F (2012) Comparing traffic discrimination policies in an agent-based next-generation network market, In: Teglio A et al (eds) Managing market complexity. Springer, Berlin/Heidelberg, pp 3–14
12. Mirza F, Beltrán F (2013) Modeling the access market of the two-sided ultra fast broadband platform in New Zealand. In: Proceedings of the 16th communications and networking simulation symposium, Society for Computer Simulation International, San Diego

13. Mingers J (2001) Combining iS research methods: towards a pluralist methodology. Inf Syst Res 12(3):240–259
14. Morton J Slowly (2013) coming to your home: fast broadband, in The New Zealand Herald, 11 March 2013
15. Mirza F, Beltrán F (2013) Using an agent-based friend circle creator model to analyze drivers of consumer choice: network effects vs. value proposition. In: Proceedings of the 16th communications and networking simulation symposium, 2013, Society for Computer Simulation International, San Diego

Influence of Losing Multi-dimensional Information in an Agent-Based Model

Sjoukje A. Osinga, Mark R. Kramer, Gert Jan Hofstede, and Adrie J.M. Beulens

Abstract This agent-based study investigates the effect of losing information on market performance of agents in a marketplace with various quality requirements. It refines an existing model on multi-dimensional information diffusion among agents in a network. The agents need to align their supply with available markets, the quality criteria of which must match the agents' information. Turnover (information entering and leaving the system) had a significant effect in the old model. Information items became obsolete based on age, causing a risk for the agents to lose valuable information. In the refined model presented here, an information item may become obsolete based on two additional aspects: (1) whether it is 'in use' for meeting the agent's current market criteria, and (2) its value, reflecting its owner's experience or skill with the information item. The research questions concern the influence of these two aspects on model outcomes. Two key parameters are *value-threshold*, below which items are candidate for disposal, and *keep-chance*, indicating the probability that *in-use* items are not disposed of.

Both simulation runs and a local sensitivity analysis were performed. Simulation results show that value-threshold is a more influential parameter than keep-chance. An interesting pattern suggesting a tipping point was observed: with increasing value-threshold, agents initially reach higher quality, but then the quality diminishes again. This pattern is consistently observed for the majority of parameter settings. An explanation is that agents with only high-valued information cannot afford to lose anything. The sensitivity analysis adds insight to where keep-chance and value-threshold are most influential, and where other parameters are responsible for observed outputs. The sensitivity analysis does not provide any further insight in why the observed tipping point occurs.

The paper also aims to highlight methodological issues with respect to refining an existing model in such a way that results of successive model versions are

---

S.A. Osinga (✉) · M.R. Kramer · G.J. Hofstede · A.J.M. Beulens
Wageningen University, Hollandseweg 1, 6706 KN, Wageningen, The Netherlands
e-mail: sjoukje.osinga@wur.nl

still comparable, and observed differences can be attributed to newly introduced changes.

1 Introduction

Autonomous producers, like farmers, make quality decisions regarding their product that must be aligned with available markets [6]. Markets distinguish themselves by means of requirements which are expressed over multiple dimensions. For example, in the case of pig farmers, markets set requirements with respect to the product (e.g. taste, leanness), price, societal concerns (e.g. animal welfare, antibiotics use), environmental concerns (e.g. carbon footprint), and so on [2]. Since these requirements are partially ordered, the farmer cannot achieve an optimal score for all quality aspects at the same time, so he must make a choice of which market to aim for.

To make profitable decisions, a farmer needs to be sensitive and responsive to information throughout the production chain [4, 5]. Information in this case represents the whole market requirements spectrum. Information can be seen as data (e.g. price information), knowledge (e.g. which breed gives certain meat characteristics, how to calculate carbon footprint), and skills (e.g. raising animals in optimal conditions, farm management). Information can disseminate through a population: farmers can exchange what they know with other farmers, agencies can educate farmers on new approaches or techniques. Although information can be shared, it does not have the same value to every owner. A farmer cannot simply copy some other farmers' knowledge or skill, but needs to build up expertise to be able to adequately use that information for his own situation. Not all farmers appreciate all information in the same way: personal preferences and differences in circumstances affect its worth [1].

Information has a lifetime, meaning that it can become obsolete. It makes sense that information loses its value over time. Some information is time related, like market information. Other information needs to be revived now and then, for example information on adopting a new technology. If nobody speaks about it anymore, then it was probably a hype that has blown over. But if it keeps going around, then the technology may be well worth considering.

At the Artificial Economics 2012 (AE2012) conference, Osinga et al. introduced the concept of multi-dimensional information in an agent-based model to align market supply with available markets [3]. The setup of this model is such that it is applicable not only to the pig farmer case, but to any situation that involves autonomous suppliers who select markets with multi-dimensional criteria and associated information requirements. The focus of the paper was on the effect of varying network structures between agents on information diffusion and market supply. Information turnover was modelled as well: new information entering the system and outdated information leaving the system. An interesting conclusion of this study was that when there is sufficient information in the system, the effect of network topology is no longer significant: markets balance out, driven by the price mechanism. Also, the effect of information turnover was very significant.

When the sheer presence of information appears to be determinant for balancing market supply, then a fair question to ask is: does it make any difference *which* information is present in the model?

2 Problem Definition

The present study is a refinement of the AE2012 model. It investigates the effect on emergent market supply of varying the conditions under which information becomes obsolete. In the AE2012 model, obsoleteness was unrelentingly determined by age, so agents ran the risk of losing information that allowed them to supply at a certain market, if it had not been renewed recently. For the present study, we investigate what difference it makes (a) when old information can be protected from becoming obsolete when it has proven its use, and (b) when young but low-valued information can become obsolete as well, instead of only old, unprotected information.

Given the assumptions of the AE2012 model, the research questions are:

1. What is the influence of *protecting useful information* from becoming obsolete on emergent market supply?
2. What is the influence of disposing of *low-valued information*, in addition to disposing of unprotected old information, on emergent market supply?

An additional, methodological purpose of this study is to describe the steps needed to refine an existing model in such a way that the results of the new model are comparable to those of the old model and differences can be attributed to the model changes.

3 AE2012 Model

A short summary of the AE2012 model is provided here. For full details see [3].

Information items are represented as triples of id, type and value. *Id* is meant to distinguish information items from one another and to indicate their age. *Type* refers to the quality dimension to which this information item belongs. This could be anything that is meaningful in the domain, e.g. health or feed in the case of pig farmers. In our model, we use abstract types A, B, C and so on. *Value* refers to the value that this information item has for its current owner, since different owners may have different knowledge or expertise to put the information item to use. For example, [24, B, 40] represents information item with id 24, of type B, which has a value of 40% to its owner. When an information item is exchanged between agents for the first time, the new owner receives a copy, but with diminished value. Information items become obsolete over time by age.

Markets represent a certain quality, and are defined as combinations of selected information types and required minimum values. Markets together cover the available quality spectrum and are partially ordered. Markets have a base price and elasticity associated with them. The price further depends on total supply, an emergent property at each time step. There is a dump market that sets no requirements, but pays nothing either.

Agents start out with a random collection of information items. At each time step, they receive one item from each network peer or from the institution (being the source of new information items that initially have maximum – expert – value). With their current set of information items, agents try to supply at a market of which they meet the requirements set, using bounded rationality. An agent has a current market, which is the last market to which he was able to supply. This implies that his current set of information items covers his current market's requirements.

4 Adjusted Model

This version of the model introduces a fourth attribute to the information items: *in-use*. This attribute is set to value True when the information item contributes to the agent's current market, and to False otherwise. Instead of disposing of old information only, the model now considers two additional criteria.

The first consideration concerns the *value* of the information item. As described above, this value reflects its owner's experience or skill with the information item. Whenever the agent receives an item with the same *id* from another agent during an information exchange event, the associated values are combined to either the same or a higher value (see for details the AE2012 paper, [3]), indicating an increase in experience and a revival of the item. This implies that low-valued information items are of relatively small use to the owner. All items below the *value-threshold*, which is a parameter of the model, are candidates for disposal.

The second consideration concerns the protection of an information item that is *in-use*, meaning that its type is required for its owner's current market. There are now two possible causes of losing information: age (an information item can reach its expiracy date), or value (its value can become lower than the value-threshold). In both situations, when the item is in use, it can be saved from disposal by a certain *keep-chance*, another parameter. If the keep-chance is 100 %, the item will never be disposed of when it is in use. If the keep-chance is 40 %, the item runs a 60 % chance of being discarded, despite its in-use status. When *value-treshold* and keep-chance are both set to 0, the model is equivalent to the AE2012 model.

In summary: All items below the value-threshold will be disposed of, except those that are in use for the current market. The ones in use are protected by a keep-chance: the higher the keep-chance, the higher their survival rate.

Refining the mechanism for information disposal triggered a change in the order in which agents perform their actions. Figure 1 presents the new order of actions. The four actions in the darker ovals used to be part of one action 'farmer step'.

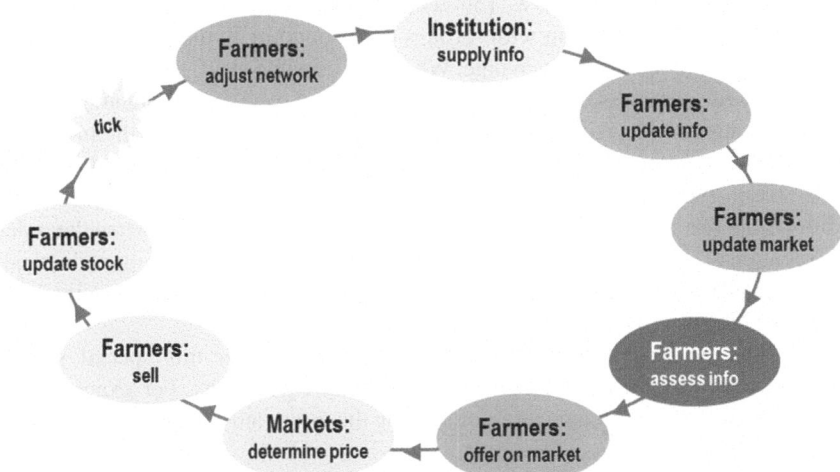

Fig. 1 A simulation time step, with farmers as suppliers. In the former AE2012 model the four darker *coloured ovals* were taken together in one 'farmer step'. The *oval* 'assess info' was added

In effect, one farmer changed his network, then updated information, updated market, and finally determined his market offer before the next farmer executed these same actions. In the new model, all farmers adjust their network, and then they all update information using the adjusted network, and so on.

The step indicated by the darkest oval 'assess info' is new relative to the AE2012 model. In this step all farmers determine which information they need to fulfil the requirements of their current market.

Another change to the model concerns the *information window*, which indicates the number of information items that an agent takes into account when determining his market options. This window is by default set to 5, meaning that for each type, the average value of the 5 highest information items is considered to see whether market requirements are met.

5 Methodological Issues

Compared with the AE2012 model, the new model required a number of changes. Some mechanisms involve drawing random numbers, e.g. using keep-chance to determine whether an information item becomes obsolete. But most new mechanisms can be set to reproduce exactly identical results to a model without that mechanism, e.g. by setting the value-threshold to zero.

To ensure that the new model outcomes would be comparable to the old outcomes, and different results could be fully attributed to the model changes and not to unintended other effects, we followed a strict procedure. After each change to the model, the model was run with settings corresponding to the previous version, to verify that outcomes did not change. Whenever possible at all, the outcomes were checked by file comparison to determine whether they were exactly identical.

Special attention has to be paid to the role of random numbers in the model. In all model runs, we set the seed for the random number generator explicitly, if only to be able to reproduce interesting results. When introducing new random effects, results cannot be identical because of these random effects. But also when the order of execution of steps changes, the effects of random numbers are no longer identical. Therefore, we concentrated all (relatively few) changes of these two types into one model development step.

All other development steps – before as well as after this special step – maintained strictly identical results for corresponding settings. For example, we could already add the mechanism for recording the in-use attribute without affecting model outcomes.

For the single model development step that influenced random numbers, we verified that model outputs were statistically equivalent as follows. Even with the same random seed, outcomes were not strictly identical. Instead, we statistically compared outcomes of multiple runs of the model versions immediately before and after this step, to verify that no essential changes were introduced inadvertently.

6 Simulation Results

For the simulations, a base case was defined in which the already present model parameters were fixed to a combination that yielded results representative for the AE2012 experiments. Only the parameters of study of the new model, value-threshold (thr) and keep-chance (kch), were systematically changed. The base case was set up with 100 farmers in a static network of 2 neighbours each and a dynamic network of an average number of friends (nfr) of 1; an information supply rate (isr) of 50, and a market set consisting of 8 markets with randomly increasing quality requirements on 4 different information types. (Referring to the AE2012 model, the network was *ring10d* and the market set *rand-inc*). With a thr of 0 and a kch of 0, the base case corresponds to the AE2012 model behaviour.

The elements to be varied were thr and kch. After evaluating some test-runs, thr was set to values 0–80 in increments of 10; kch was set to 0, 30, 60, 80, 90, 100. All combinations of these parameters were repeated ten times to mitigate the effects of randomness. The result figures below each show one run, with agents active on dump, low, medium and high market segments, accumulated over time.

Figure 2 shows one of the runs with kch 0 and increasing thr's. The top line, thr 0, reflects the behaviour of the AE2012 model (the base case). The kch is 0, meaning that information items that are in use have no special protection, which allows us to see the influence of thr. We observe that a higher thr means a considerable shift in market balance. With a thr from 0 up to about 40, we observe that the market share

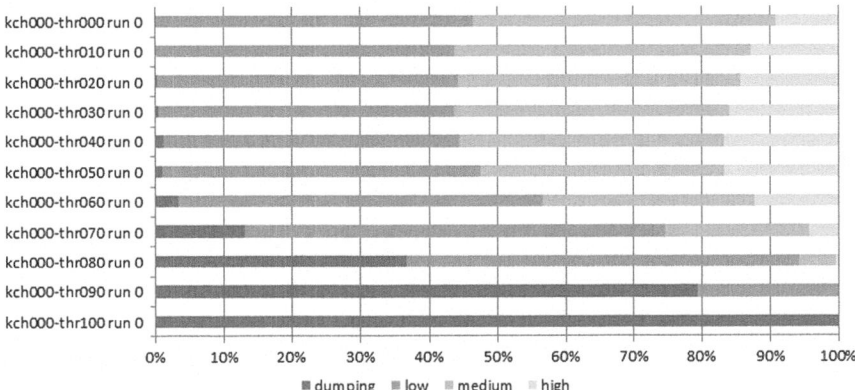

Fig. 2 Result of one simulation run with kch = 0 and varied thr. The *bars* indicate the percentages of farmers active on dump, low, medium and high market segments, accumulated over time

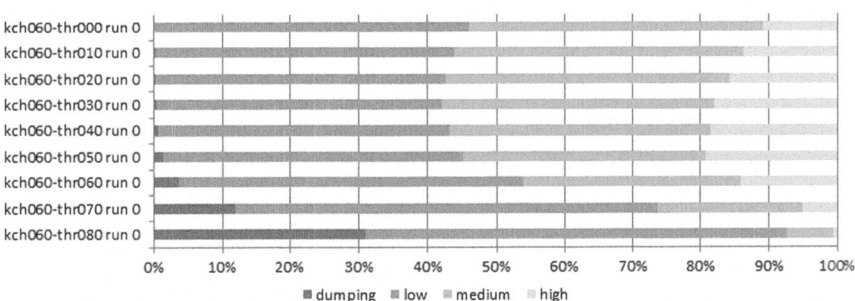

Fig. 3 Similar to Fig. 2, but now with kch = 60 and thr varied up to 80. The same pattern of increasing and then decreasing high market segments is visible

with highest quality increases. But when thr increases further, we observe that the high market share decreases again, the low market share gains ground, and agents even have to take resort to the dump market. We can explain this phenomenon as follows: when thr is low, any information item that is disposed of has a low value, which will not hurt the agent very much. But when thr is high, only precious high-valued items remain, and whatever item is disposed of will be a loss to the agent. Without it, he may not be able to maintain the requirements of his high quality market anymore. We consistently observe this pattern of initially improving and subsequently losing quality with all values for kch tested, except value 100.

Figure 3 is another example result where kch was set to a fixed value (60 this time), and thr was varied up to 80. We see that the typical pattern of increasing and decreasing high quality segments is clearly visible again. In comparison with Fig. 2, we observe that a higher kch protects information slightly better so that the reached quality level is higher.

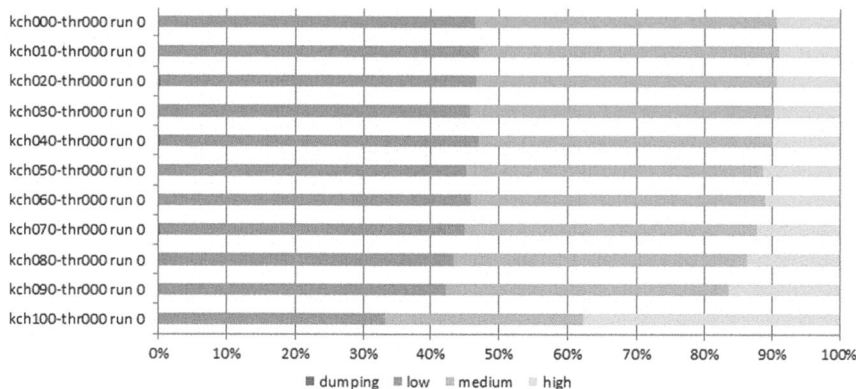

Fig. 4 Result of one simulation run with thr = 0 and varied kch. The *bars* indicate the percentages of farmers active on dump, low, medium and high market segments, accumulated over time

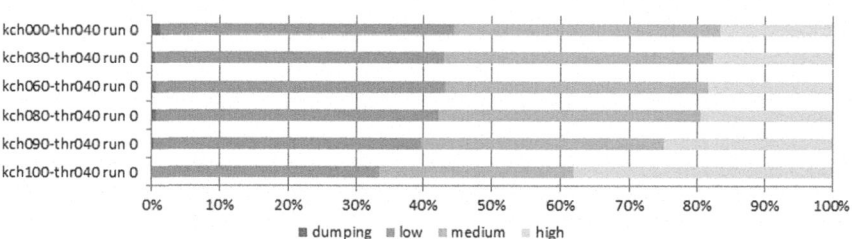

Fig. 5 Similar to Fig. 4, but now with thr = 40 and varied over less kch's. The higher market segments are larger, compared with thr 0, because disposed items are the low-valued ones

Figure 4 shows results for varying kch with a constant thr of 0. Again, the top line reflects the base case. With thr 0, age rather than low value is a reason for disposal, which allows us to see the influence of kch. A higher kch increasingly protects the agent from losing information that he currently has in use, enabling him to maintain or improve his current market. Only when kch is 100, the agent can keep a successful set of information items forever, and we see indeed that high quality market shares are relatively large. In situations where kch is lower and his useful information items become of old age, the information items that an agent holds will not be sufficient to maintain his quality anymore. We observe that a lower kch means that agents are less able to supply at high quality markets, and that medium and low quality markets gain in market share.

Figures 5 and 6 show similar results as Fig. 4, but now with thresholds set to a fixed value of 40 and 80, respectively, and with kch varied over less values. For thr of 40 (Fig. 5), we observe that agents reach higher quality segments than with thr of 0. This makes sense, because the low-valued items are disposed of and high-valued items are the ones that remain. However, when thr is 80 (Fig. 6), this effect turns against itself. There are no low-valued items left to dispose of, and loss of any

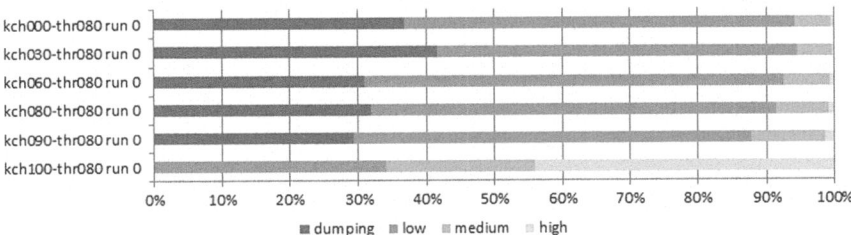

Fig. 6 Similar to Fig. 5, but now with thr = 80. Except for kch = 100, when there is no disposal at all, quality is much lower than with thr 40. When thr is high, only high-valued items remain, and disposal of any item is a loss to the agent

item is a problem. As a result, agents cannot maintain the high qualities anymore and even have to resort to the dump market. Only when kch is 100, and all items are protected, agents can still supply at high quality markets.

7 Sensitivity Analysis

We performed a local sensitivity analysis in order to see how sensitive to parameter changes our results are – and especially the apparent tipping point for varying kch visible in Figs. 2 and 3. We varied the parameters shown in Table 1 one at a time, with the values shown there. Only kch and thr were varied both one at a time and together. For one at the time variation, thr was set to 0 and kch from 0 to 100 in steps of 10, and the other way around. When they were varied together, their values were set to all combinations of 0, 30, 60 and 90. Each run of a particular parameter set was repeated ten times. The mean result of these runs is reported as outcome.

The parameters in the list of Table 1 are the number of farmers in the simulation (farmers), the information supply rate (isr), the average number of friends that a farmer has (nfr), the number of time steps that the simulation lasted (steps) and the information window that farmers have (window).

The sensitivity of the parameters to the outcomes were expressed according to the formula shown in Eq. 1 and normalized according to Eq. 2:

$$sensitivity = \frac{v_{hi} - v_{lo}}{p_{hi} - p_{lo}} \qquad (1)$$

$$normalized\ sensitivity = \frac{v_{hi} - v_{lo}}{p_{hi} - p_{lo}} * \frac{p_b}{v_b} \qquad (2)$$

In these equations, v represents the mean value for the variable over all runs (output), and p represents the parameter that was varied (input). The subscript b

Table 1 Settings chosen for local sensitivity analysis

Parameters	Low	Base	High
Farmers	90	100	110
Isr	45	50	55
Nfr	0	1	2
Steps	1,000	2,000	3,000
Window	4	5	6
Kch			
Thr		see text	

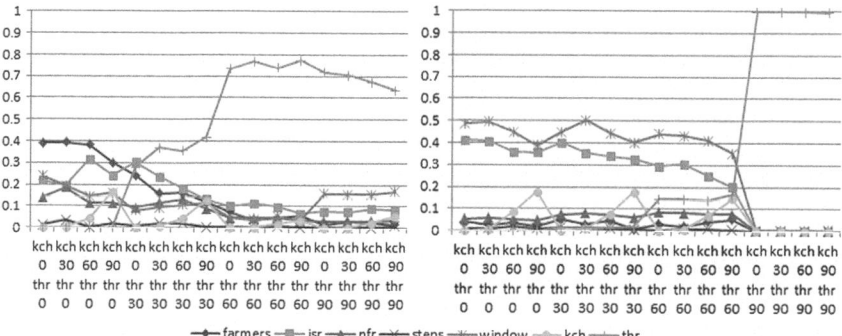

Fig. 7 Relative sensitivity of parameters varied according to Table 1; kch and thr varied simultaneously. *Left*: dump market; *right*: high markets

stands for the value of the base case. Subscripts *lo* and *hi* indicate the lower and higher than base case values for the parameter, respectively.

The results of the sensitivity analysis are summarized in Figs. 7 and 8. Figure 7 shows the combined influence of kch and thr, when varied together. The left panel shows the relative influence of all parameters under consideration on the dump market share. The right panel shows this influence on the high market share. For conciseness reasons, we do not show the relative parameter influences on low and medium market shares.

We observe that the thr parameter has a relatively larger influence than kch. For the dump market, a thr from value 30 onwards is very dominant. We see that other parameters are responsible when thr is not yet so dominant, of which initially the number of farmers seems most influential, and also the information supply rate (isr). The farmer influence is negative, meaning that when the number of farmers increases, fewer farmers resort to the dump market. With our current analyses we cannot explain why this happens.

For the high markets, we see that the thr parameter is dominant from value 90 onwards. When thr has value 90, all information items below 90 are thrown away, so effectively nothing happens anymore. As a consequence, no other parameter has any influence. Two other parameters have a relatively high influence as long as thr is not yet dominant. These are the information supply rate (isr) and the information

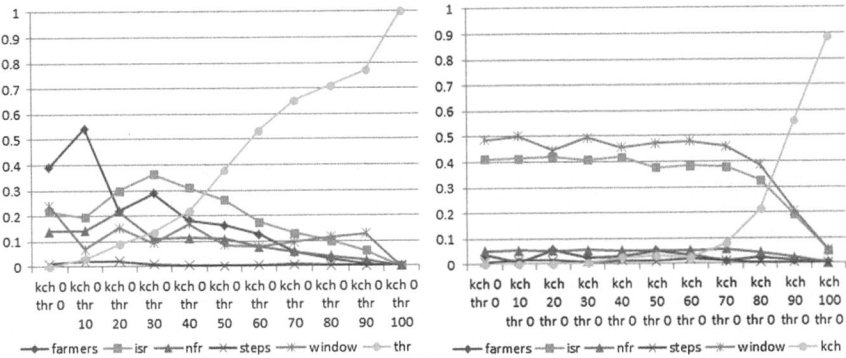

Fig. 8 Relative sensitivity of parameters varied according to Table 1; kch and thr varied one at a time. *Left*: kch = 0, thr varied (dump market); *right*: thr = 0, kch varied (high markets)

window. This makes sense: the more information in the system, the more high quality markets come within reach (as was also an AE2012 conclusion). And for window: when we vary the number of information items that are taken into account during decision making, this is expected to be especially sensitive for high quality requirements. It also makes sense that the number of steps, the number of farmers, and the number of friends do not have a significant influence (the latter also being an AE2012 conclusion).

The left panel in Fig. 8 shows the results of local sensitivity analysis with a kch of 0 and a varying thr, for the dump market segment. From the absolute results, shown in Fig. 2, we know that farmers start to enter the dump markets when thr value is about 40 or higher. That is consistent with what we see here: from about value 50, the influence of thr is increasing and becomes very dominant. It is indeed the treshold value that is responsible for farmers having to supply at the dump market.

The right panel in Fig. 8 shows similar results, but now for the high market segment, and with thr set to 0 so that the influence of kch can be observed. For the major part of the graph, with low and medium kchs, the parameters isr and window are dominant. For values higher than 70, kch becomes an influential factor for the high markets. In the absolute results, shown in Fig. 4, we saw that the high market segment is increasing with kch. With the sensitivity analysis results, we can add the insight that kch is only responsible for this when it has a value higher than 70. Below that value, parameters isr and window mainly determine whether farmers are able to supply at high markets.

8 Conclusions and Discussion

In this paper, we reconsidered the way agents dispose of their information in our model. Next to expiracy because of age, an information item can now also be disposed because of too low value, indicated by a *value-threshold*. An information item can receive the special status of *in-use* that indicates whether the information item is needed for an agent to keep his current market. The *keep-chance* indicates the probability that in-use items will be saved from disposal. We ran simulations where we focussed on the value-threshold and keep-chance parameters. We also performed a sensitivity analysis of the results. Coming back to our research questions, of which question 1 referred to the keep-chance and question 2 to the value-threshold, we can conclude that the influence of value-threshold is very significant, and the influence of keep-chance is moderate.

Sensitivity analysis shows that the influence of most other model parameters is according to expectation. Parameters isr and nfr behave according to the AE2012 model outcomes. The information window has an influence where it makes sense. The number of steps in the simulation has almost no influence. Sensitivity of networks and markets is difficult to establish, since they are not numerical, but those were tested and reported in our AE2012 model. The only deviation is the number of farmers, which seems to have an influence, but for the explanation of which we need further investigations.

The sensitivity analysis results add insight to the simulation results, with respect to the relative influence of value-threshold and keepchance. Sometimes their influence is quite absolute, but sometimes other parameters are more influential. The sensitivity analysis results do not give any cause to interpretation conflicts. The simulation results show a consistent pattern where a tipping point is suggested. However, the sensitivity analysis gives no decisive explanation for or insight in this phenomenon.

An additional goal of this paper was to highlight some methodological aspects, especially concerning the procedure for guarding comparability of results from successive model versions. This is a particular concern in agent-based models, where the order in which agents execute their actions requires careful attention.

References

1. Arens L, Plumeyer CH, Theuvsen L (2012) Determinants of the use of information: an empirical study of german pig farmers. Int Food Agribus Manag Rev 15(1):51–72
2. Bonneau M, Lebret B (2010) Production systems and influence on eating quality of pork. Meat Sci 84(2): 293–300
3. Osinga SA, Kramer MR, Hofstede GJ, Beulens AJM (2012) Multi-dimensional information diffusion and balancing market supply: an agent-based approach. In: Teglio A, Alfarano S, Camacho-Cuena E, Gins-Vilar M (eds) Managing market complexity: the approach of artificial economics, vol 622. Springer, Heidelberg, pp 184–194
4. Verbeke W (2001) The emerging role of traceability and information in demand-oriented livestock production. Outlook Agric 30(4):249–255
5. Verdouw CN, Beulens AJM, Trienekens JH, Van der Vorst JGAJ (2011) A framework for modelling business processes in demand-driven supply chains. Prod Plan Control 22(4): 365–388
6. Wever M, Wognum N, Trienekens J, Omta O (2010) Alignment between chain quality management and chain governance in EU pork supply chains: a transaction-cost-economics perspective. Meat Sci 84(2):228–237

Least Susceptible Networks to Systemic Risk

Ryota Zamami, Hiroshi Sato, and Akira Namatame

Abstract There is empirical evidence that as the connectivity of a network increases, there is an increase in the network performance, but at the same time, there is an increase in the chance of risk contagion which is extremely large. If external shocks or excess loads at some agents are propagated to the other connected agents due to failure, the domino effects often come with disastrous consequences. In this paper, we design the least susceptible network to systemic risk. We consider the threshold-based cascade model, proposed by Watts [13]. We propose the network design model in which the associated adjacency matrix has the largest maximum eigenvalue. The topology of such a network is characterized as a core-periphery structures that consists of a partial complete graph of hub nodes and stub nodes that are connected to one of the hub nodes. The introduced network can reduce the turbulence of shocks triggered and prevent the spread of systemic risk. By both mathematical analysis and agent-based simulations, we show that the slightly differences of the structure of network causes systemic risk.

1 Introduction

How do we keep the performance of networked systems under accidents or failures? The ongoing progress of networking in essential utilities (ex. power grid, transportation, Internet) bring significant benefits to the quality of our life, but the networked system hold a certain danger that a failure of only single node in the networked system can wipe out all other nodes. This phenomenon is widely recognized after the Great Northeast Blackout in the U.S. 1965. It was a result of a

---

R. Zamami (✉) · H. Sato · A. Namatame
Department of computer science, National Defense Academy of Japan, Yokosuka, Japan
e-mail: em51049@nda.ac.jp

sequence of recursive failures, and this kind of failures is called "cascade failure". There are many studies to find an efficient treatment [1,5–7,16], because the impact of the failure is far more severe than previously thought of the direct impact by the failure of nodes [10].

Watts studied the relationship between average degree of a network and robustness of each node (it is called as "threshold" of a node in the paper) [4, 13]. He found the cascade failure could occur in the specified region in terms of average degree (it is named "cascade window".) when all nodes have same robustness. When the network is too sparse (there are few links) to be connected, the impact cannot spread via nodes. When the network is sufficiently dense (there are many links), the size of the impact of failures becomes relatively small because each node has enough number of other nodes to transmit or receive physical quantities. Motter studied overflow model and showed the redistribution of the load of flow cause the cascade failures [10]. In many man-made networks, each node (ex. the packet router in the Internet) has few extra capacity (or resources) to respond to accidents (ex. a traffic burst due to the failures of other routers). Their results show that the load turbulence triggered by the failure of only single node can cause catastrophic damage of the network.

These studies about cascade failures imply the important fact that the robustness of each element in the network does not show the robustness of the entire network. A network topology defines where and which nodes communicate or interact with others. In the case of a certain network, a part of failures of a network can spread via vulnerable parts of a network or change the balance of the load drastically. Then, we have to take into account not only the performance of each element but also the network topology to design a robust network to cascade failures.

Several papers use models for cascade failures to understand the dynamics mathematically and numerically. Here we consider two above-mentioned models (threshold model and overflow model). Threshold model can model social risks including rumor, crime, riot and other phenomena including the threshold dynamics based on the fraction of the state of the adjacent nodes. The each node becomes failed when the fraction of failed neighboring nodes exceeds the own threshold. At the specific condition, a failure of single node can change the state of adjacent nodes sequentially and the impact may spread through the network. On the dynamics, hub nodes (nodes with many links) are least susceptible to the external failures, while the stub nodes (they have a few links and connect to single hub node) are susceptible to it. The problem of network design on the threshold model is how hub nodes are interconnected for a robust network under the restriction of the resources. More properly, we have to find better probabilistic degree distribution $P(k)$, which defines the density of nodes with k links, for the more robust network.

The overflow model can model the congestion problem in networked system including Internet, transportation, and power grid. This model sets a capacity and a load on each node initially. The removal of some nodes by accidents or failures usually causes redistribution of flow of physical quantities. Then, the impact of initial failures can appear at not only adjacent nodes but also distant nodes. In the overflow model, in which the flow diffuse on the shortest path, hub nodes will be selected as relaying points to decrease the hop distance. Then the failures of hub nodes will cause the major redistribution of flows. To design the robust network on the overflow model, it is intuitive idea that the homogeneous network is best, because all nodes have sufficient alternative adjacent nodes after node failures. However, the homogeneous network has side effects for the network performance that the average hop distance between nodes prone to be large due to the lack of hub nodes.

In this paper, we show least susceptible network topology against cascade failures based on both the threshold model and the overflow model, and introduce a simple method to build the network. The network consists of hub nodes and stub nodes. The hub nodes make complete graph to prevent the spread of failures by threshold dynamics, and stub nodes, which are belongs to only singe hub node, are useful to reduce the size of load turbulence on the overflow model. Interestingly, the maximum eigenvalue of an adjacency matrix of the network is close agreement with mathematically upper limit value. We also show the robustness of the network comes from the hub-cluster, and it does not depend on the topology of stub nodes. By numerical simulations, we show the cascade failures do not occur on the networks and the benefit of hub nodes makes average hop distance to be shorter than scale free networks.

2 Cascade Models

2.1 A Threshold Model

Here, we introduce the threshold model as one of models for cascade failures (the recursive sequence of failures) [4,8,11,13,14,17,18]. The model consider a failure of a node propagate when the fraction of neighboring failed nodes exceeds a threshold of a corresponding node. Let the networked system have N nodes and M links and each node i has the state si = 0 (healthy or not failed) or $s_i = 1$ (unhealthy or failed). The state s of nodes depends on the fraction of failed nodes in neighboring nodes. We assume all nodes have same threshold ϕ and simultaneously update their own sate following Eq. 1) at each discrete time.

$$s_i = \begin{cases} 1 \sum_{j \in N_i} s_j/d_i \geq \phi \\ 0 \sum_{j \in N_i} s_j/d_i \leq \phi \end{cases} \quad (1)$$

where s_j and N_i represents the state of node j and adjacent nodes of node i respectively.

Granovetter originally introduce the threshold model for the several human activities including the spread of riots, rumor and crime [4]. The model is expanded as a model of unifying systemic risk where "risk" diffuses by threshold dynamics [9].

The model and the traditional percolation model seem very similar and have many common features, because in both models the change of node state propagates based on the external sate in the network [12, 15]. The differences from the percolation model are that the threshold model uses the fraction instead of absolute numbers and it is studied on the complex network, while the percolation is usually studied on the lattice space.

The probability that the number of neighbors d_i of each node is k follows the degree distribution P(k). If the P(k) follows Poisson distribution, almost all nodes have same number of neighbors, while a few nodes have huge number of neighbors if it follows Power law distribution. The degree of node d_i also represents the tolerance of node i against external failures. In the case of nodes with small degree, they may be severely impacted by the failure of only single adjacent node, while in the case of hub nodes, they are very stable because they have many other nodes even if one of them becomes failed.

D.J. Watts et al. studied the relationship between the average degree z of the network and the threshold ϕ against cascading failures. They found that there is a region in which a small amount (one or two nodes) of failures can make the sequence of node failures and it spreads to the entire network. They called the region "cascade window". They also showed a mathematical tool to specify the cascade window theoretically from the degree distribution P(k) of the network,

$$\sum_{k=0}^{k_*} k(k-1)P(k) = z, K_* = \lfloor 1/\phi \rfloor \qquad (2)$$

To understand the effect of the difference of the degree distribution P(k), we use three kinds of distribution (Homogeneous P(k=z)=1: all nodes have same degree z and they are interconnected randomly by those links, Poisson distribution (RND) $P(k) \sim e^{-z}z^k/k!$: all nodes have almost same number of links z [3], Power law distribution (SF(BA)) $P(k) \sim k^{-\gamma}$ with $\gamma \sim 3$: there are many nodes with a few links and a few nodes with many links (called hub nodes). We obtain the degree distribution by using Barabási-Albert model [2].).

Figure 1 shows the theoretical cascade window of networks from those degree distributions P(k). The area bounded by line, which is labeled "Cascade", represents cascade window. In the area, the cascade window can occur by a few of failed nodes. The size of area represents how susceptible to cascade failures a corresponding network is. We can see that the Homogeneous network is most robust against cascading failures, while SF (BA) network is most susceptible. Then, we can get the relationship about the robustness R of the network against cascade failures in terms of the size of cascade window,

$$R_{Homogeneous} > R_{RND} > R_{SF(BA)} \qquad (3)$$

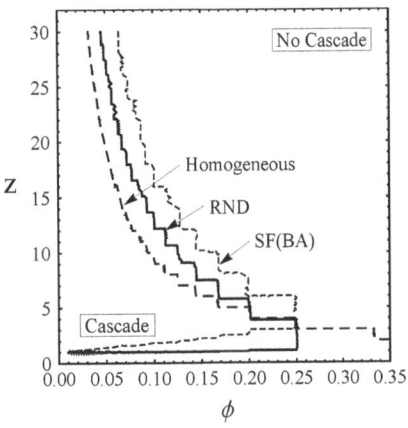

Fig. 1 The theoretical cascade windows of several networks (*SF(BA)*): Scale-free network by Barabási-Albert model, *RND*: Random network, Homogeneous: Random regular network): The cascade failure can expand to the entire network in the region bounded by each line

3 Network Topology with Largest Maximum Eigenvalue

3.1 Motivation

We can classify nodes into two categories (hub nodes and stub nodes) in terms of the size of degree. In the case of the threshold model, the hub nodes have two functions as a brake and an accelerator of the cascade failures. When a stub node fails, a stable hub node prevents the spread of the impact (as brake). However, a hub node spreads the failure to many other nodes (as accelerator), when a hub node itself fails.

In the case of the overflow model, the hub nodes are usually selected as a pathway to reduce the average hop distance between nodes and carry large amount of flows as relaying point to the destination. The failure of them causes major changes in the network connectivity and the balances of loads. Then the existence of hub nodes can be reason of cascade failure.

If the impact of the failures of hub nodes can be reduced drastically by the network topology, the network should be least susceptible to cascade failures and have the sufficient performance in terms of average hop distance. Here we introduce a heuristic method to make least susceptible network to cascade failures based on the threshold model and the overflow model, which has both hub and stub nodes.

3.1.1 A Network Design Model

Our model consists of only two steps to build a network, as shown in Fig. 2. Let us consider we have N nodes and M links for a network. At first, we make the n-complete graph as a core of a network, in which n-nodes are connected each other completely and has $n-1$ links. Each node needs at least one link to connect itself to other node. Then, we can have the relationship about the number

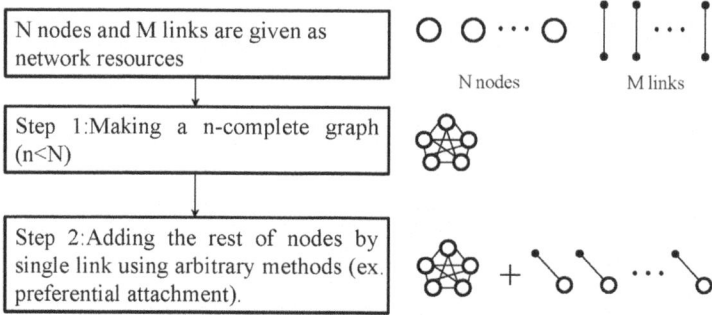

Fig. 2 The Procedure of building proposed networks

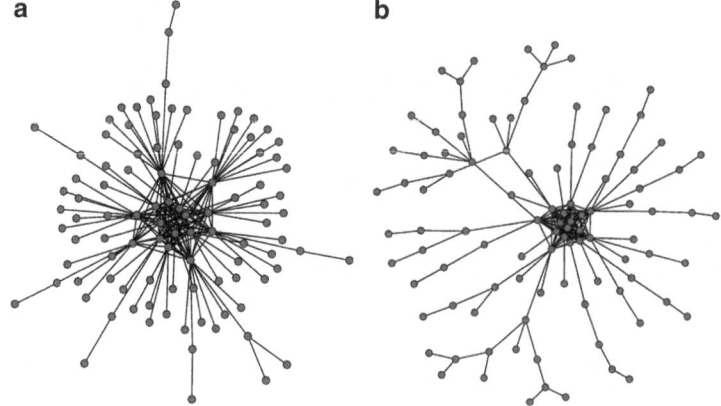

Fig. 3 The least susceptible networks: the networks have 100 nodes and 200 links. The number of nodes in the center core is 16. (**a**) the network that consists of complete graph and preferentially attached nodes (CPA). (**b**) The network that consists of complete graph and randomly attached nodes (CRA)

of links as Eq. 6. By isolating n from Eq. 6, we can get Eq. 7, which defines the size of parameter n. Secondly, the rest of nodes are attached to the core network by using arbitrary methods (ex. preferential attachment like BA model [2] or random attachment). Figure 3 shows the examples of our models, which have 100 nodes and 200 links. All stub nodes belongs to only single hub nodes and they connects each other via the center core cluster. The CPA networks, which consists of the core cluster and preferentially attached nodes, have minimum average hop distance compared with scale free networks by Barabási-Albert model [2], random network, and homogeneous network when the average degree is less than ten (see Fig. 4). This implies the introduced network (CPA) has good connectivity and the index is usually used to estimate the network performance [1, 10]. The introduced networks also have discriminative value of the maximum eigenvalue of the adjacency matrix, which characterize the network topology. The upper limit of the maximum eigenvalue is

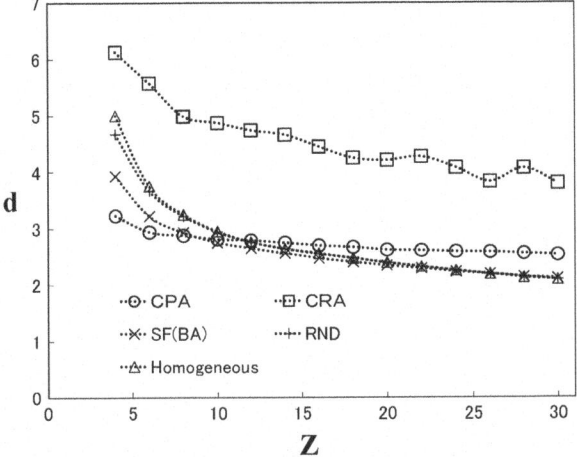

Fig. 4 Relationship average hop distance d and average degree z: Each network has 500 nodes. Each represents same network in Figs. 1 and 3

defined mathematically as shown in Eq. 6 [19]. We calculate the eigenvalue and compare it with the eigenvalue of the theoretical upper limit and the other networks. Interestingly, the introduced networks (CPA, CRA) have the eigenvalue that almost equals to the upper limit along the increasing of the average degree z (see Fig. 5).

$$M = \frac{n(n-1)}{2} + N - n \qquad (4)$$

$$n = \lfloor \frac{3 + \sqrt{9 + 8(M-N)}}{2} \rfloor \qquad (5)$$

where the symbol $\lfloor \cdot \rfloor$ represents the floor function Eq. 5.

$$\lambda_1 \leq \sqrt{1 + N(Z-1)} \qquad (6)$$

4 Simulation on Cascade Failure

We do numerical simulations to show the robustness of our proposed networks (CPA, CRA) against cascade failures.

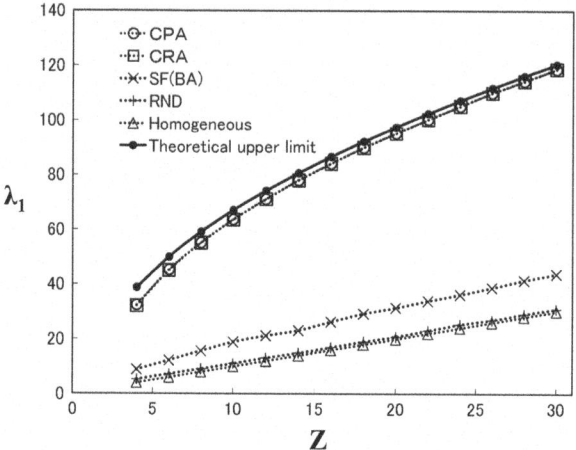

Fig. 5 Relationship between maximum eigenvalue of adjacency matrix λ_1 and average degree z Each network has 500 nodes. Each legend represents same network n Figs. 1 and 3

4.1 Simulation Settings

We compare the robustness of several networks including the introduced networks (CPA, CRA) in Sect. 3, scale free networks by Barabási-Albert model [2], random networks by ER model [3] and homogeneous (random regular) network. All networks have 500 nodes.

At first, we compare the theoretical cascade window of networks and simulate cascade failures based on the threshold model. Initially almost all nodes are healthy ($s_i = 0, \forall i$), and all nodes have same threshold value. At the time t = 0, single node, which is selected randomly at each time, becomes failed as trigger of cascade failure. We obtain the boundary condition (cascade window) for the cascade failures in terms of average degree and threshold value. The boundary obtained from the results of 1,000 tests.

Second, we simulate cascade failures based on the overflow model. We allocate capacity to each node, which is based on the node betweeness. We remove a node that has maximum node betweeness as a trigger of cascade failures. After that, the load is recalculated and some nodes become failed when the newly assigned load exceeds capacity of them. We observe the size of largest connected component and the average hop distance between nodes in it.

4.2 Results: Threshold Model

Figure 6 shows the theoretical cascade window of each network from degree distribution P(k), as a function of the average degree z and the threshold value of

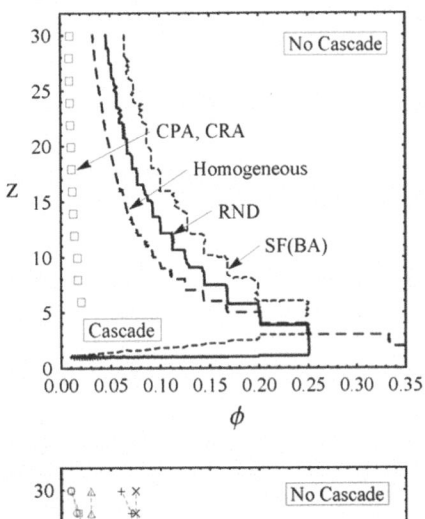

Fig. 6 Comparison of the theoretical cascade window between proposed networks (CPA, CRA) and other networks: Each legend represents same network in Figs. 1 and 3

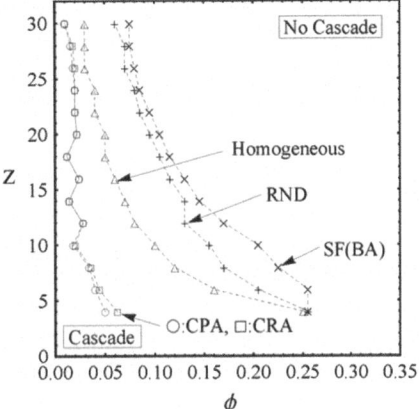

Fig. 7 Comparison of the cascade window by numerical simulations: Each network has 500 nodes. Each legend represents same network in Figs. 1 and 3

each node ϕ. We can see that homogeneous network has a relatively small cascade window but our proposed networks have smaller one. This result is in agreement with intuition, because all paths between nodes go through the cluster of hub nodes in CPA or CRA network, which is very stable against the external failures. In the case of other networks, there are some paths without hub nodes, along which the impact of the failure can reach many other nodes. Even if the hub node is selected as a trigger in CPA or CRA network, the hub node will change only its stub nodes. The complete graph of hub nodes can work like a firewall against cascade failure.

Figure 7 shows the numerical simulation results. We can see that the scale free network (SF(BA)) is actually susceptible to cascade failures but our proposed networks are not susceptible. For instance, in the case average degree $z = 4$, the CPA and CRA have five times of robustness compared with SF(BA). As shown in Fig. 3, the CPA and CRA have different network topology except the center complete graph, but they have similar cascade window. This implies the robustness of the network comes from the center cluster.

4.3 Simulation Settings on Systemic Risk

We assume that an individual node in the network has assets, include External assets and Interbank borrowing, denoted by E and B, and that a node has a liabilities include Interbank loan, Deposit and Capital buffer, denoted by L, D and C. and we define a θ as percentage of interbank borrowing to total assets. and γ as percentage of capital buffer to total assets. We then construct balance sheets by following aspects.

- *Interbank Borrowing and Interbank Loan.* Each link expresses financial relations. For node k, out-degree k_{out} is borrowing asset, B_k. and in-degree k_{in} expresses Interbank loan, L_k. and size of these portion, B_k and L_k equal to each degree, k_{out} and k_{in}.
- *Total Assets and External Assets.* Interbank assets of node k, I_k in its total assets TA_k to be θ that is, $I_k = \theta TA_k$, and then $TA_k = I_k/\theta$ And $E_k = (1-\theta)I_k$.
- *Capital Buffer.* Capital buffer of node k in its total assets, TA_k to be γ that is, $C_k = \gamma TA_k$.
- *Deposit.* Liabilities include interbank loan, Deposit and Capital buffer. Already, we got interbank loan and capital buffer. therefore D_k equal to $TA_k - (L_k + C_k)$. According to above aspects, we complete the balance sheets.

We simulate the process of shock propagation as interaction of bank. We modeled interbank market by muti-agent. For bank k in interbank market affected by initial shocks, S_k. We assume shock to be first absorbed by its capital buffer. then its interbank liabilities to be followed by its deposits, as the ultimate sink. If shock is bigger than capital buffer, bank k defaults and a surplus shock, $S_k - C_k$ is divided up equally and transmitted to its creditor banks. however, a supremum of transmit shock is L_k. The transmission continues until the all shocks are absorbed. And then, we measure the systemic risk each networks with count defaulted nodes without a node affected by initial shock.

4.4 Results: Systemic Risk

In this paper, we will consider balance sheets comprising 45% of interbank borrowing, 3% of capital buffer to total assets. And, we paid attention to both network, CPA and CRA. We show that CPA and CRA network are hard to cascade failures. We simulate these networks, 500 nodes, $\langle k \rangle = 20$ (undirected), These quantities are the result of 500 runs, and are reported in Figs. 8 and 9.

These figures shows that the structure of periphery nodes make large differences to systemic risk. Periphery nodes link together makes the network fragile. These network has a same maximum eigenvalue, also same cascade window but slightly difference of the structure of network changes the systemic risk.

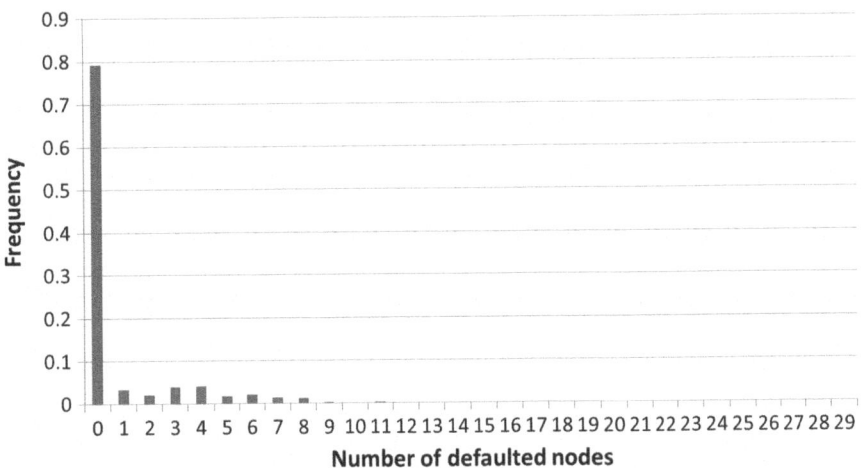

Fig. 8 Number of defaulted nodes vs. frequency (CPA network), cascade failure could not spread in most cases

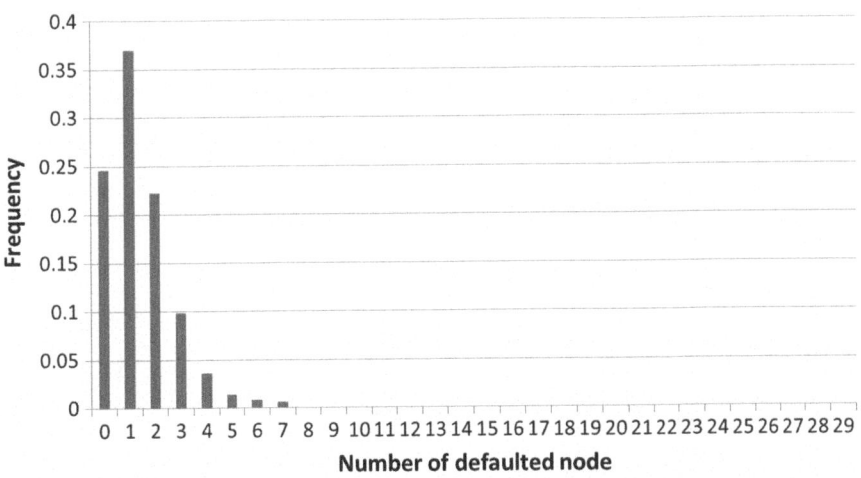

Fig. 9 Number of defaulted nodes vs. frequency (CRA network). Several cases, cascade failure spread into interbank market. A case that up to seven nodes went defaults occurred

5 Summary

We have studied the robustness of the network against cascade failures based on the threshold model and the over flow model and introduced least susceptible network to the cascade failures. The network consists of hub nodes and periphery nodes that are belong to one of hub nodes in the cluster. One hub node and periphery nodes make a module. The network consists of many the modules, and a center cluster (complete graph) of hub nodes unifies them. The results of numerical simulations have shown that the network topology can drastically reduce the damage of the cascade failures. The center cluster of hub nodes contributes to prevent the spread of failures in the case of cascade failures based on the threshold model, because each hub node in the cluster has sufficient redundancy (many adjacent hub nodes), to be robust for the external failures. The core-periphery architecture of the network is useful to suppress the turbulence of load by failures of nodes in the case of cascade failures based on the overflow model, because the load of each hub node is equal to the amount of flows that start or end at its own periphery nodes. But in case of shock propagation, these networks has almost same cascade window but has entirely different about systemic risk. The structure of periphery nodes is causes of disparity of systemic risk.

References

1. Albert R (2000) Errror and attack tolerance of complex networ. Nature 406(27):378–382
2. Barabási A, Albert R, Jeong H (2000) Scale-free characterixtics of random networks: the topology of the World-Wide Web. Phys A 281:69–77
3. Erdös P, Rény A (1959) On random graphs. Publ Math (6):290–297
4. Granovetter M (1978) Threshold models of collective behavior. Am J Soc 83(6):1420–1443
5. Gutfaind A (2010) Optimizing topological cascade resilience based on the structre of terrorist networks. PLos ONE 5(11):e13488
6. Komatsu T, Namatame A (2011) Dynamic diffusion processes in evolutionary optimized networks. Int J Bio Inspir Comput 3:384–392
7. Komtasu T, Namatame A (2011) An evolutionary optimal network desighn to mitigate risk contagion. In: Proceeding of IEEE international conference natural computation (ICNC), Shanghai, vol 4, pp 1980–1985
8. López-Pintado D (2006) Contagion and coordination in random networks. Int J Game Theory 34(3):371–381
9. Lorenz J, Battistion S, Schweitzer F (2009) Systemic risk in a unifying framework for cacading processes on networks. Eur Phys J B Condens Matter Complex Syst 71:441–460
10. Motter AE, Lai YC (2002) Cascade-based attacks on complex networks. Phys Rev E 66(6):065102
11. Schelling TC (1973) Hockey helmets, concealed weapons, and daylight saving: a study of binary choices with externalities. J Confl Resolut 17(3):381–428
12. Stauffer D, Aharony A (1991) Introduction to percolation theory. Taylor and Francis, London
13. Watts D (2002) A simple model of global cascades on random networks. PNAS 99(9): 5766–5771
14. Watts D (2007) Influentials, networks, and public opinion formation. J Consum Res 34: 441–458
15. Wolframs S (1983) Statistical mechanism of cellular automata. Rev Mod Phys 55:501–644
16. Xu J, Wang XF (2005) Cacading failures in scale-free coupled map lattices. Phys A 349: 685–692
17. Young P (2009) Innovation diffusion in heterogeneous populations: contagion, social influence, and social learning. Am Econ Rev 99(5):1899–1924
18. Young P (2010) The dynamics of social innovation. PNAS 108(9)21285–21291
19. Yuan H (1988) A bound on the spectral radius of graphs. Linear Algebra Appl, 108:135–139. http://dx.doi.org/10.1016/0024-3795(88)90183-8

The manufacturer's authorised representative in the EU is Springer Nature Customer Service Centre GmbH, Europaplatz 3, 69115 Heidelberg, Germany. If you have any concerns regarding our products, please contact ProductSafety@springernature.com

Printed and bound by CPI Group (UK) Ltd, Croydon, CR0 4YY

23/03/2026

02076664-0007